D0875940

MAN'S RELATION TO THE UNIVERSE

In 1969, the Faculty of Natural Sciences and Mathematics at the State University of New York at Buffalo established the Distinguished Visiting Lectureship of the Faculty of Natural Sciences and Mathematics. This lectureship is sponsored by the State University of New York at Buffalo and awarded annually to an outstanding scientist, who presents a series of four lectures at the University. These lectures are to cover as wide a range of material as possible, not only recent developments in research, but also the more philosophical aspects of particular areas of endeavor. The lectures are also intended to stimulate and encourage an interest in the sciences in as wide an audience as possible, among both the University and the Buffalo communities. This monograph is based on the 1973 lectures of Sir Bernard Lovell, F.R.S.

Distinguished Visiting Lecturers of the Faculty of Natural Sciences and Mathematics at the State University of New York at Buffalo

1970 Linus Pauling

1971 Fred Hoyle

1972 George Wald

1973 Bernard Lovell

MAN'S RELATION TO THE UNIVERSE

Bernard Lovell
UNIVERSITY OF MANCHESTER

W. H. Freeman and Company
San Francisco

Cover: Drawing of Halley's comet as shown on the Bayeux Tapestry.
(Courtesy of Yerkes Observatory.)

Library of Congress Cataloging in Publication Data

Lovell, Alfred Charles Bernard, Sir, 1913–
Man's relation to the universe.
Includes bibliographical references.
1. Astronomy. I. Title.
QB44.2.L68 520 75–14096
ISBN 0–7167–0356–4

Printed in the United States of America

9 8 7 6 5 4 3 2 1

PREFACE

THIS BOOK IS AN EXTENSION of four lectures that I delivered in the Distinguished Visiting Lecturer Series at the State University of New York at Buffalo in October 1973. The title and the main content of the book are the same as the lectures. However, the first lecture, delivered under the title of *International Relations and Economics of Space Exploration*, has been somewhat changed to deal more specifically with the practice and cost of the purely astronomical investigations reported in this book. The wider issues of space exploration, its origins, economics, and civil and military applications, which I also referred to in my original lecture, have been dealt with more fully in a small book published in the fall of 1973.*

From the terrestrial concern with equipment and its cost the lectures moved to a consideration of new concepts of the nature of the universe derived from observations with contemporary apparatus, either ground-based or in space. Finally, I considered the immense conceptual difficulties that arise when we discuss our relation to the universe, especially with regard to the recent evidence in favor of a singular high-density beginning. The subject matter of these lectures is at

*B. Lovell, *The Origins and International Economics of Space Exploration,* Edinburgh University Press and John Wiley & Sons Inc., New York, 1973.

the forefront of some of the most exciting avenues of contemporary research, and where appropriate I have included results obtained since the lectures were delivered.

I am grateful to Chairman Dr. Michael Ram and to the other members of the Lectureship Committee for the invitation to deliver these lectures. To them and to many other members of the State University of New York at Buffalo I express my gratitude for providing the stimulating atmosphere in which these lectures were delivered.

Bernard Lovell

Jodrell Bank
February 1975

CONTENTS

MAN'S RELATION TO THE UNIVERSE

CHAPTER ONE

TECHNIQUES AND ECONOMICS OF ASTRONOMICAL INVESTIGATION

THE PRESENT PERIOD in the development of astronomy is typified by the intensive investigation of two topics—the physical condition of the solar system and the nature of the remote regions of time and space. These investigations are stimulated by two intellectual passions common to all ages, the desire to understand how the Earth came into existence and the attempt of the human mind to understand the significance of time and space and the evolution of the universe. In recent years there has been a new feeling of excitement and missionary zeal for these researches, because our age has witnessed the development of new techniques for astronomical research.

TECHNIQUES OF OBSERVATION

From the beginning of the 17th century, when Galileo first used a small telescope to study the heavens, through the three and a half centuries to the end of World War II, it seemed that the only instrument for gaining astronomical knowledge was the optical telescope. Further, as manufacturing techniques improved, astronomers benefited from the ability to make ever larger telescopes. From the 1-inch telescope of Galileo they progressed through the 48-inch telescope of Sir William Herschel (1789) and the 72-inch instrument of Lord Rosse at Birr (1845). In the first half of this century a telescope with a 100-inch mirror was built on Mt. Wilson in California; then in 1948 the Hale telescope on Palomar Mountain was commissioned, with a mirror 200 inches in diameter. Because these telescope mirrors collected more light than their predecessors they were able to penetrate farther into space than ever before, and their resolving power was superior to that of any previous instrument.

Subsequent to the discovery by Fraunhofer in 1814 of the lines in the solar spectrum and, shortly afterwards, in the spectra of the stars, the spectroscope became an important auxiliary instrument for use with the optical telescope. The first tentative applications of photography to astronomy were made in 1840. The first dry plates were used in 1876, and from that time improvements in photographic techniques led to significant increases in the sensitivity of telescopes. The optical telescope, assisted by photographic and spectroscopic techniques, has made immense contributions to our knowledge of the universe. Indeed, researches with the Mt. Wilson telescopes have led to our present conception of the general structure of the universe. It is easy to understand the belief of astronomers that the future of observational astronomy would be governed by optical telescopes of great size. The temperatures of stars had been found to be in the region of thousands of degrees, and hot bodies at such temperatures were known, from fundamental physical principles, to emit the peak of their energy in the visible or near-visible region of the spectrum (see Figure 1.1); in any case the Earth's atmosphere would obscure any radiation emitted at wavelengths

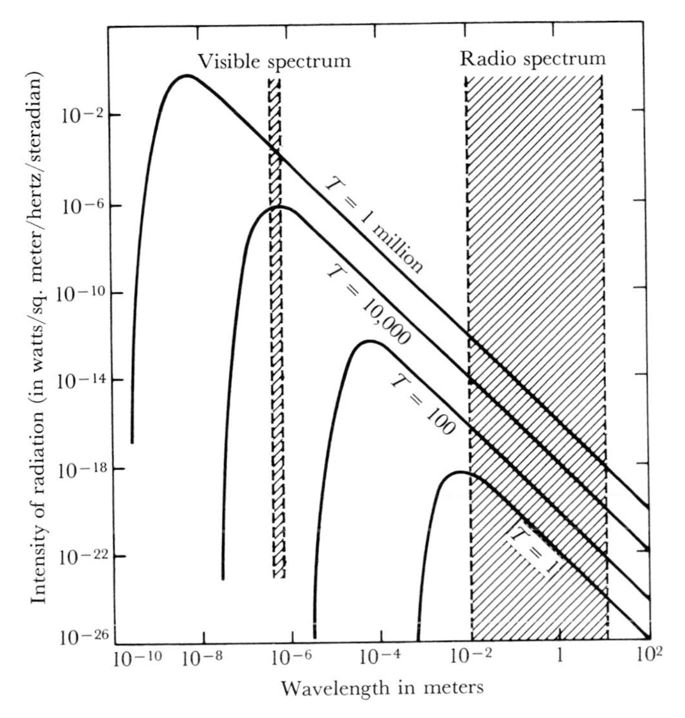

FIGURE 1.1. *The intensity of radiation plotted against the wavelength for bodies of various temperatures T of 1 K, 100 K, 10,000 K and 1,000,000 K. The regions of the visible spectrum and the radio spectrum are indicated by the dashed lines.* (*From B. Lovell,* Our Present Knowledge of the Universe, *Manchester University Press, 1967.*)

shorter than the ultraviolet or longer than the far red, as shown in Figure 1.2.

The remarkable fact that during the last 25 years observational astronomy has not developed in this way has at least two important and distinct reasons. The first is Karl Jansky's discovery, in 1931, of the existence of long-wave radio emissions from space. Until the immediate postwar years this discovery was almost disregarded by astronomers. Then a sequence of discoveries in quick succession, using the new techniques of radio astronomy, caused a complete

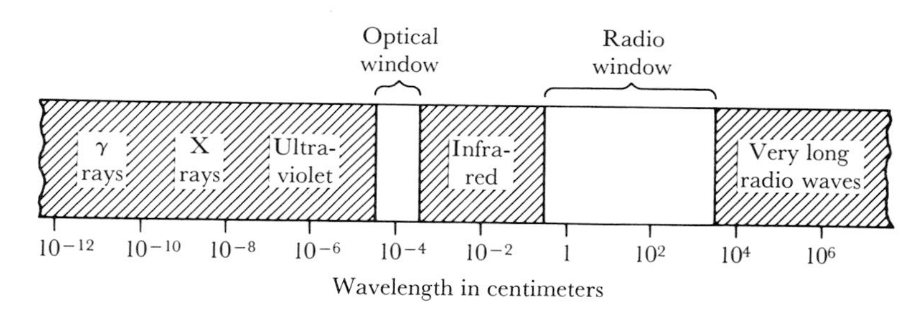

FIGURE 1.2. *The electromagnetic spectrum, showing the regions of transparency of the Earth's atmosphere and ionosphere. (From B. Lovell,* The Exploration of Outer Space, *Oxford University Press, 1962.)*

reorientation of views. It soon became clear that fundamental processes are operating in the universe that give rise to strong emissions at long wavelengths, well beyond the peaks in Figure 1.1 classically associated with the emission from a hot body. Many important contemporary discoveries derive from the application of these radio techniques to the study of the heavens.

Second, since 1957 there has been the impact of the Earth satellite and space probe on astronomical research. Astronomers had little concern with the initiation of space research. The International Union of Radio Science and the International Union of Geodesy and Geophysics were the two major scientific groups to urge the development of these techniques for studies of the atmosphere, ionosphere, and nearby space during the 1957–1958 International Geophysical Year. On the other hand, many astronomers of international renown were strongly opposed to the new developments of space research, because they felt that large telescopes on Earth, which could be built at a cost small compared with that of space vehicles, were the correct development in astronomical research. Even after so short a time, and in spite of the small part of the space budget so far devoted to astronomical research, it can be seen that this opposition made a bad misjudgment of the potential of the Earth satellite and space probe.

Within space research there are two developments that are proving to be as significant as the earlier development of radio astronomy. The belief that the emission from bodies in the universe could be represented as in Figure 1.1 has been shown by the radio telescope to be erroneous at the long-wave end. Equipment for recording X-ray and γ-ray emissions, carried in satellites above the obscuring layers of the Earth's atmosphere and ionosphere, have similarly showed that hitherto unsuspected processes operate to produce short-wave high-energy emission in these wave bands. In fact, modern astronomical techniques have opened the entire electromagnetic spectrum to study over a wavelength range of at least 10^{13} to 1, from the 10^{-12} meters of the γ-rays to more than 10 meters at the long radio-wave end of the spectrum. Any modern attempts to understand the universe must seek to interpret emissions over this great range of wavelengths. The second development within space research of great significance to astronomy has been the success in sending scientific instruments to the Moon and the planets and, in the case of the Moon, the ability to recover physical samples either by manned or unmanned excursions.

THE OPTICAL TELESCOPE

A striking feature of the contemporary scene in optical astronomy is the increasing doubt among the new generation of optical astronomers that the future of this aspect of astronomy lies with instruments even larger than the 200-inch Palomar telescope. The reasons for this doubt are complex. First, there is the immense difficulty and cost of constructing a telescope larger than the 200-inch and of finding a site on Earth where the seeing conditions and the background light of the sky would allow its potential to be realized. The Russians have attempted to do this. They have built, in the northern Caucasus Mountains, a telescope with a mirror diameter of 236 inches. It is difficult to obtain precise information from Russian astronomers about the status of this project, although there seem to be well-attested reports that there have been difficulties with the mirror.* As to the cost, my information from a reliable source is that the entire project, which includes the mountain access and buildings on the site, has cost 200 million rubles ($400 million). This is at least an order of magnitude more costly than any telescope so far built—a straightforward replacement of the 200-inch Hale telescope is now estimated to cost $25 million. There are also fundamental reasons for doubting whether such projects are now justified. The 200-inch telescope can photograph stars ten million times fainter than can be seen by the naked eye, and 20 times fainter than the darkest parts of the night sky. Hence the problem in penetrating to fainter objects is essentially one of detecting the contrast between starlight and the surrounding sky light so that, from this size, the performance of the telescope would improve only slowly with increase in size, whereas the cost would probably increase as the cube of the size.

There are two further important considerations. The first is that the photographic plate is an inefficient means of recording—about 99 percent of the photons that fall on it are wasted—and it has threshold and saturation levels. The deficiencies of the plate have stimulated the development of electronic means of recording the image. The use of photomultipliers 25 times more efficient than the photographic plate was a major advance, but they could only view a single resolution element of the image at a time. Now further advances are in hand that combine these high efficiencies with the ability to photograph the whole two-dimensional image at the same time. The idealized system now in prospect reads the data directly into a computer, so that the astronomers can obtain the image from the computer store at any time and optimize the exposure. These electronic recording techniques already have a major impact on telescope size, since a 40-inch with electronic instrumentation is equivalent to the 200-inch with photographic plate. Similarly, the 200-inch with this instrumentation could be regarded as equivalent in sensitivity to a 1,000-inch using photographic plate recording;

*An illustrated account of a visit in September 1973 to this telescope near Zelenchukskaya has been given by A. G. Davis Philip in *Sky and Telescope* 47, 290, 1974.

and the estimates for building a 1,000-inch telescope are at least $2 billion.

Given these possibilities for major improvements in telescopes of relatively small size, there still remains the urge to achieve even greater sensitivity, and hence to apply these new recording techniques to telescopes of the largest possible size. The formidable cost, today, of building telescopes with 200-inch and larger mirrors has stimulated the investigation of optical telescope arrays. The idea is to achieve a large collecting area either by mounting a number of small mirrors on a common frame or by combining them as separate telescopes. The concept of an optical telescope array should now be realizable because of the precise automatic control of telescope positions possible with the use of computers, and because of the possibility of integrating the separate images obtained with electro-optical recording techniques.

This transformation of outlook in optical astronomy in recent years has occurred abruptly. In 1964 the National Academy of Sciences published the report of a committee under the chairmanship of A. E. Whitford, recommending a ten-year program for U.S. ground-based astronomy.* The committee made firm recommendations that three large telescopes be built, including the projected 150-inch Kitt Peak instrument and two of 200-inch diameter. There was also a firm recommendation that engineering studies be made to determine the cost and feasibility of building a 400-inch, or possibly a 600-inch, telescope. Although the inefficiency of the photographic plate was stressed, the report remarked on the subject of image tubes that "the bright prospect held out by the considerably higher quantum efficiency of the photoelectric cathode relative to the photographic plate has not been realized after several years' effort. In only a few limited, special cases results not attainable by photography have been achieved." The cost of the program, including a number of smaller telescopes, was placed at about $68 million, of which the three large telescopes were estimated at $60

**Ground-based astronomy. A ten-year program,* National Academy of Sciences, Washington, D.C., 1964.

million. Of this program the Kitt Peak telescope, with an aperture of 158 inches and a cost of about $10 million, was dedicated in June 1973.*

Eight years after the Whitford report the National Academy of Sciences issued the report of a committee under the chairmanship of Jesse L. Greenstein.† In this report all mention of proceeding with studies of a 400- to 600-inch telescope has vanished, to be replaced by emphasis on the revolution in electro-optical recording techniques. In a projected budget of $83 million, the report recommends the expenditure of $15 million for the provision of the major existing American telescopes with advanced sensors and controls, and then states, "An operating multimirror telescope equivalent to a 150- to 200-inch single mirror is estimated to cost about $5 million. Further funding up to $25 million should then be provided to build the largest possible telescope within that budget—either a multiple-mirror one with an effective aperture of 400- to 600-inches, if the concept proves feasible, or a conventional 200-inch telescope." At the present time (1974) the Smithsonian Astrophysical Observatory and the University of Arizona are jointly constructing a telescope consisting of six 72-inch mirrors mounted on a common frame, which will use a laser technique for bringing the six separate images to a common focus. It is calculated that by this means the light-gathering power equivalent to a 170-inch telescope will be achieved, but at a fraction of the cost of a single-mirror telescope of that aperture.

This sudden transformation in outlook, caused by the rapid development of electro-optical and computer techniques in the few years after the publication of the Whitford report, now makes it extremely unlikely that the Soviet precedent of building a telescope larger than the 200-inch will be followed elsewhere. The advantage of light-gathering power will be realized by improvements in the

---

*A telescope of similar size, constructed as a joint Anglo-Australian enterprise and to be operated jointly, was dedicated in October 1974 at Siding Spring, New South Wales. The cost was $A16 million.

†*Astronomy and astrophysics for the 1970's,* National Academy of Sciences, Washington, D.C., 1972.

efficiency of the sensors and by the construction of arrays of smaller telescopes—probably in the 60- to 80-inch category. Furthermore, because of turbulence in the Earth's atmosphere, it seems unlikely that any site on Earth can be found that will enable the resolving power of large telescopes to be realized. Except in certain specialized measurements, turbulence limits the angular resolution of optical telescopes to the order of 0.5 seconds of arc. That is, the 200-inch, which has a theoretical angular resolution of a few hundredths of a second of arc, does little better under normal conditions than a telescope one-tenth the size.

ASTRONOMY FROM SPACE VEHICLES

The serious handicap presented by the atmosphere to the effective operation of large optical telescopes on Earth has stimulated the consideration of orbiting optical telescopes. If a telescope could be operated from a space platform above the atmosphere there would be three significant advantages.

(*a*) The theoretical resolving power could be realized. That is, a 120-inch telescope on Earth has a resolving power (about 0.5 seconds of arc) no better than can be obtained with a 12-inch telescope. In space, the resolving power is diffraction-limited, or ten times better.

(*b*) The sensitivity of the largest Earthbound telescope now in use, the 200-inch Hale telescope, is fundamentally limited by the fact that it has to record an image against the light of the background sky and, because of the turbulence, over a disc of diameter about 0.5 seconds of arc. In space the equivalent disc is determined by diffraction limitations and, for a 120-inch, would be about 0.05 seconds of arc in diameter. There is a difference of 100 times in the interfering area of sky involved. Hence, for similar sensor systems a 120-inch in space would be 5 magnitudes more sensitive than the 200-inch on Earth. That is, an object just within the sensitivity limit of the 200-inch could be seen at ten times the distance by the 120-inch telescope in space.

(*c*) The atmospheric optical window of Figure 1.2 is limited to a range of wavelengths from about 2,900 Å at the short-wave, violet end to 10,000 Å at the long-wave, red end. A major contribution to the long-wave limitation is absorption by the water vapor in the atmosphere. Measurements of the infrared radiation from the heavens can be made from high and dry mountain sites, or by observing from balloons or aircraft that carry the instruments above the tropopause. On the other hand the short-wave limitation, which cuts off the ultraviolet light, is caused by absorption by ozone at a height of 25 to 40 kilometers in the atmosphere, and at shorter wavelengths (in the 1,000- to 2,000-Å region) by absorption by ionized oxygen molecules in the atmosphere. Thus astronomical observations in this ultraviolet domain can be carried out only by using instruments in rockets or space vehicles above the entire atmosphere.

Astronomers have long realized that spectroscopic information in the ultraviolet region and measurements of the general shape of the spectrum would be of great significance in the study of the Sun, the stars, and the extragalactic nebulae. Since a space telescope would be diffraction-limited, there would be the additional bonus of increased resolving power, for observations in the ultraviolet, by at least a factor of 2 over the figures given in (*a*) and (*b*).

The importance of such ultraviolet observations from space has been recognized by the National Aeronautics and Space Administration, and in spite of some unfortunate early failures, since 1967 a number of satellites have been operating successfully, carrying small telescopes for studies of the ultraviolet radiation from the Sun and celestial objects. In June of 1973 a press release from NASA gave details of the international team of scientists selected to define the experiments to be carried on a large space telescope (LST) of 120-inch aperture, which is hoped to be launched by the Space Shuttle in the 1980's.* The 1972 report of the Greenstein committee

*As envisaged at present the LST with its 120-inch mirror will have a focal length of 30 feet. Weighing about 10 tons, the instrument will orbit at an altitude of 648 to 778 kilometers and the guidance system will be capable of holding on to a target for extended periods to within 0.005 seconds of arc.

expressed concern at the hiatus in the plans for further telescopes in space after the completion of the current series of orbiting solar and astronomical satellites. Their strong recommendation was that $15 million per year be allocated to a continuation of this program, and $35 million per year for a program leading to the LST; and that if this could not be launched until the mid-1980's, then an intermediate 60-inch space-telescope project should be initiated.*

Although investigations of the ultraviolet emissions from the stars and other objects can be made from space vehicles, another fundamental limit sets in at a wavelength of 912 Å because of absorption by the interstellar hydrogen. The window opens again at 100 Å in the soft X-ray region. Until 1962 no one expected that the study of radiations in the X-ray region would be of importance to our knowledge of the universe, yet in the subsequent decade it has become evident that X-ray observations from space vehicles may well be as surprising and as important to our knowledge of the universe as the radio observations have been.

Because of absorption in the Earth's atmosphere, observations of X rays can be made only at heights greater than 100 kilometers. On 18 June 1962 an Aerobee rocket launched from White Sands, New Mexico, carried a geiger counter system designed by Giacconi and others of the American Science and Engineering laboratories and Rossi of MIT.† The equipment was intended to study fluorescence X rays produced on the lunar surface by X rays from the

*The Greenstein committee estimated the overall cost of the LST at $1,000 million. The 1974 NASA estimates are that the LST could be built for far less than this amount if the Space Shuttle were used for the launching instead of an expendable booster. The NASA estimate is of the order of $500 million for the LST launched with the Shuttle in the 1980's. The present (1974) status is that the LST project is threatened with major setbacks because of economic pressure on NASA. The House committee with control over the space-agency appropriation voted to delete the $6.2 million requested in the current fiscal year to finance further definition and technical studies of the LST. In the event a ceiling of $3 million was agreed, provided NASA found the money elsewhere in its budget of $3,200 million, and Congress directed NASA to consider a later launch of a less expensive telescope, with financial assistance from other countries.

†R. Giacconi, H. Gursky, F. R. Paolini, and B. Rossi, *Physical Review Letters* 9, 439, 1962.

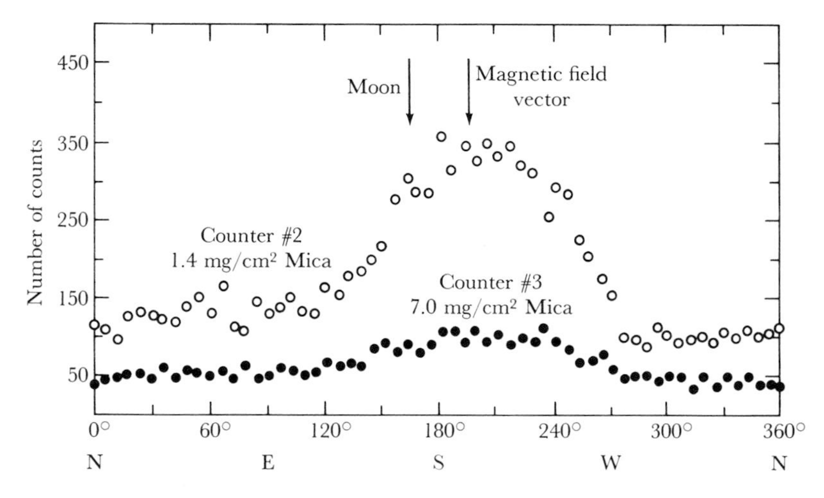

FIGURE 1.3. *The first record of X rays from the Milky Way. The ordinate is the number of counts accumulated in 350 seconds in each 6-degree angular interval. The abscissa is the azimuth angle. It was deduced that the peak was caused by a source emitting X rays at a wavelength of 3 Å situated about 10 degrees above the horizon and 10 degrees in width, and that this source was in the direction of the galactic center. (From the paper by R. Giacconi, H. Gursky, F. R. Paolini, and B. Rossi,* Physical Review Letters 9, *439, 1962.)*

Sun. The rocket reached an altitude of 225 kilometers and was above 80 kilometers for 350 seconds. The system was sensitive to radiation in the 2- to 8-Å wavelength region, and it was expected that the lunar fluorescence X rays would give an intensity of about 0.1 to 1 photons per square centimeter per second. In fact, the record of Figure 1.3 was obtained, showing a peak intensity at 3 Å of 5 photons per square centimeter per second. The equipment received in a cone of 5 degrees, and from the attitude of the rocket it was concluded that the observed peak was caused by X-ray emission from the direction of the galactic center.

Less than a year after this discovery, scientists of the Naval Research Laboratory obtained more detailed information from apparatus carried in another Aerobee rocket launched from White

Sands on 29 April 1963.* This equipment was sensitive in the wavelength range 1 to 8 Å. The field of view was 10 degrees, but because of the roll of the rocket nearly the whole sky was scanned. The results revealed the existence of two localized sources of X-ray emission, one in Scorpius, the other in the constellation of Taurus near the Crab nebula. In addition, a diffuse background of X rays was apparent.

These remarkable results stimulated interest in X-ray observations, and in the following years many rocket-borne flights were made for the recording of X rays from space. Notable among these was the rocket flight of July 1964, from which the group from the Naval Research Laboratory succeeded in observing the occultation of the Crab nebula by the Moon, thereby establishing that the X-ray source in Taurus found in the 1963 flight was indeed coincident with the Crab nebula.†

Until late in the year 1970 these X-ray observations from rocket-borne equipment continued, although because of the short time for which a rocket was at a high enough altitude, the total recording in all of these flights was not more than 10 hours. From this work it became evident that a considerable number of X-ray sources existed in space, as well as a diffuse background. Naturally, great interest attached to the possible identification of these X-ray sources with previously known objects.

On 12 December 1970 the entire subject of X-ray astronomy was revolutionized by the launching of the UHURU X-ray satellite. Giacconi and his team at American Science and Engineering during the first 70 days of observing catalogued 120 X-ray sources, about 80 of which were believed to lie within the galaxy. The catalogue published in 1974 listed 161 X-ray sources located in the analysis of 125 days of data.‡ More than half of these sources (about 95) lie

---

*S. Bower, E. T. Byram, T. A. Chubb, and H. Friedman, *Nature 201,* 1307, 1964.

†S. Bower, E. T. Byram, T. A. Chubb, and H. Friedman, *Science 146,* 912, 1964.

‡R. Giacconi, S. Murray, H. Gursky, E. Kellogg, E. Schreier, T. Matilsky, D. Koch, and H. Tananbaum, ASE-3249, *Astrophysical Journal Supplement 27,* 37, 1974.

at low galactic latitudes and are believed to be objects within the Milky Way system. The remainder are at high galactic latitudes and are believed to be extragalactic, although many sources in both categories remain to be identified.

On 21 August 1972 the Copernicus satellite was placed in orbit. This contained a 32-inch-aperture ultraviolet telescope from Princeton University and an X-ray telescope designed by R. L. F. Boyd and his colleagues of University College, London. This X-ray equipment is now adding significantly to the data already obtained by the UHURU satellite.

These pioneering X-ray investigations from space vehicles have been followed by several other Earth satellites carrying more sophisticated equipment. The latest (1974) of these are the Astronomical Netherlands Satellite (ANS), launched on 30 August 1974, and the British UK5 (Ariel 5), launched in October 1974. A few years ago NASA had an ambitious project known as the High Energy Astronomical Observatory (HEAO). The plan was to launch two large satellites in 1975 and 1977, followed by two further launchings by the Space Shuttle in the 1980's. Each satellite would have carried 12,000 pounds of equipment for the study of X rays, γ rays, and cosmic rays. In 1973 the estimated cost of each of these satellites was $275 million. Because of financial retrenchment NASA was forced to abandon this important project, and the plan was revised to provide for three payloads of only about one-fifth the weight. The first of these, HEAO-A, planned for 1976, will carry X-ray equipment nearly ten times more sensitive than that in the UHURU satellite. The second, HEAO-B, planned for 1978, will carry even more sensitive X-ray equipment and an X-ray telescope with a resolution of 2 seconds of arc.

The UHURU survey was carried out in the energy range of 2 to 10 keV (10 keV is equivalent to a wavelength of about 1 Å) and the majority of the results discussed above refer to rocket-borne or satellite results in this energy range (wavelengths between 1 and 10 Å), that is, in the soft X-ray spectrum. A number of sources have been detected in the hard region of the X-ray spectrum, in the energy range from 20 to 500 keV (down to wavelengths of about 0.02 Å). Sensitivity is more difficult to obtain at these high energies, but because of the greater penetrating power of hard X rays, many of the results in this region have been obtained from apparatus carried in high-flying balloons.

The investigation of γ rays from space presents great difficulties, and the development of γ-ray astronomy has not yet made the spectacular advance of X-ray astronomy. The incident flux of γ rays from space is not only very low (at 100 MeV, for example, it is only of the order of 10^{-5} photons per square centimeter per second), but it must be investigated against a background radiation of charged cosmic-ray particles that is 10,000 times greater. The first indication of γ rays from space was obtained in 1962, when equipment in the Ranger 3 Moon probe recorded radiation with energy in the 1-MeV range, apparently of extragalactic origin. Subsequently, γ-ray recording equipment has been used in balloons, and in 1968 the OSO-3 satellite carried recording equipment for the 100-MeV region. The TD-1 satellite carried equipment for investigations up to 25 MeV, and SAS-B up to 50 MeV. During the trans-Earth coast of Apollo 15, 4–7 August 1971, γ-ray observations in the energy range 0.3 to 27 MeV were made by equipment carried in the service module. For very much higher energies, in the 10^5 to 10^7 MeV range, the intensities are probably too low to be detected by instruments capable of being carried into space in the foreseeable future, but in these ranges ground-based detectors can be used to observe the Cerenkov radiation from the cosmic-ray showers produced in the atmosphere by the primary γ ray.

The most surprising feature of the γ-ray observations so far has been the detection between 1969 and 1972, by the Vela satellites operated by the U.S. Department of Defense, of 16 short bursts of γ rays.* The four Vela spacecraft involved were equally spaced in a circular orbit at distances of 10^5 kilometers from the Earth. The bursts varied in duration from less than 0.1 seconds to 30 seconds, and the directional properties of the apparatus made it

---

*R. W. Klebesadel, I. B. Strong, and R. A. Olson, *Astrophysical Journal Letters 182,* L85, 1973; also *Astrophysical Journal Letters 188,* L1, 1974.

possible to exclude the Earth and the Sun as possible sources. The IMP-6 satellite launched on 14 March 1971 contained a γ-ray monitoring system, and subsequent analysis revealed that this system had also detected the bursts of γ rays. The sources of these bursts, and whether they are galactic or extragalactic, is not yet known.

THE RADIO TELESCOPE

The paper published in 1932 by Karl Jansky included results that are now recognized to be historic in the annals of astronomy. Since 1930 he had been studying the direction of arrival of atmospherics on a wavelength of about 15 meters, which was in use for communication purposes. The aerial that he constructed at the Bell Telephone Laboratories in New Jersey was a simple array of dipole rods, about 30 meters long and 4 meters high. The framework of this aerial was mounted on four wheels (from a Model T Ford) and a motor drive rotated the device one revolution in about 20 minutes. Retrospectively, it has to be regarded as the world's first radio telescope, for in addition to the terrestrial sources of atmospherics that were his primary concern, Jansky observed a component of the noise that appeared to emanate from a fixed direction in the sky. His conclusion was unambiguous because he made the simple, but fundamental, observation that with his aerial fixed in a given direction, the maximum of the signal occurred with a period of 23 hours 56 minutes. This is a sidereal day—the period of the Earth's rotation with respect to the stars.

It is an extraordinary feature of the history of science in the 20th century that the supreme importance of Jansky's discovery was not immediately recognized. It is often remarked that his discovery attracted no attention, but this was not, in fact, the case. The *New York Times* of 5 May 1933 contained a front-page full-column report of the discovery, and it was featured in radio programs. This makes it even more surprising that no professional astronomers or academic institutions pursued the investigations. Jansky himself wanted to pursue the research, and proposed that a 100-foot-diameter dish-shaped aerial be built. But the Bell Telephone Laboratories did not agree that this expense was worthwhile, and since, as far as their commercial purposes were concerned, Jansky had fulfilled the objectives of the investigation, he was not allowed to pursue work on this new discovery.* The cost of Jansky's equipment must have been small, and certainly a minor item in the budget of the Bell Telephone Research Laboratories.

Although the news of Jansky's discovery did not stimulate the professional scientific institutions or the astronomers to activity, it did inspire an amateur investigator. Grote Reber was a young graduate radio engineer; as a hobby, and at his own expense, he built in the yard of his home at Wheaton, Illinois, a dish reflector 31 feet in diameter. The reflector was made of galvanized iron sheet secured to a wooden framework cut to a parabolic shape. It was mounted so that it could be steered in elevation only—the Earth's rotation providing the other motion that enabled the telescope to scan the heavens. He built the device in four months of 1937, and it cost him $1,300. Reber used a much shorter wavelength than Jansky—1.87 meters—and with his 31-foot-diameter aerial he received the radio emissions in a narrower beam (about 12°). In 1940 Reber published the first detailed radio map of the heavens, and at last the professional astronomers showed interest in his work. They concluded that the strength of the radio waves showed a correlation with the distribution of the stars, but that individual stars were not detectable as radio emitters. Reber suggested that the radio waves were produced in ionized interstellar hydrogen gas in the galaxy.

A valuable history of these early researches and of the subsequent developments has been given by J. S. Hey,† himself a pioneer of

---

*Jansky continued to work at the Bell Telephone Laboratories, but unfortunately died at the age of 44 in 1950, when the scientific world was on the verge of understanding the immense importance of his discovery. An account of his work has been given by his brother C. M. Jansky in *Proceedings of the Institute of Radio Engineers 46*, 13, 1958.

†J. S. Hey, *The evolution of radio astronomy*, Paul Elek Scientific Books, London, 1973.

the subject, who discovered the radio outbursts from the Sun while engaged in research for the British Army during World War II. An extraordinary feature of the story is that for more than 15 years after Jansky's work the subject was not pursued by astronomers or by academic institutions, and the pioneering discoveries of the radio waves, both from the sky and from the Sun, were made without significant investment of money. Immediately after the end of World War II the subject was opened up in the academic field in England and Australia.* A few years later the Dutch began to make vigorous progress inspired by J. H. Oort of Leiden and the prediction of H. C. van de Hulst that it should be possible to detect the radio-spectral line emissions from galactic neutral hydrogen on a wavelength of 21 centimeters. Ironically, interest remained low in the United States, and in those early postwar years in America the main center of interest was at the Naval Research Laboratory in Washington, where a 50-foot steerable paraboloid was constructed on the roof of the laboratory. At last, in 1954, D. H. Menzel and B. J. Bok of Harvard, with J. B. Wiesner of MIT, initiated proposals for a national facility. The president of the Associated Universities Inc. (AUI), L. V. Berkner, was approached and plans for a National Radio Observatory at Green Bank, West Virginia, were formulated. By 1959 an 85-foot telescope was in operation, and Otto Struve became the first director.†

Whereas in the first 15 years after Jansky's discovery the embryonic science of radio astronomy cost little, the subsequent developments became expensive, by academic standards. These researches, together with nuclear physics, were the first of the "big sciences" to invade academic institutions. This was a natural consequence of the technical problems faced by radio astronomers. Radio waves have wavelengths of the order of a million times longer than the

*In England by M. Ryle at the Cavendish Laboratory, Cambridge, and by the author at Jodrell Bank (University of Manchester); in Australia by J. L. Pawsey at the Radio-Physics Laboratory of the Commonwealth Scientific and Industrial Research Organisation (CSIRO).

†In 1962 he was succeeded by D. S. Heeschen. A recent account of the observatory has been given by J. W. Findlay, *Sky and Telescope 48,* 352, 1974.

radiation in the optical spectrum. A 10-inch optical telescope has a resolving power of about 0.5 seconds of arc. Because of the greater wavelength a radio telescope would have to be many miles in diameter to achieve this kind of resolution. The steerable paraboloidal form of radio telescope has been used extensively, but clearly dish sizes of this order are impossible. For many years, the fully steerable 250-foot-aperture radio telescope at Jodrell Bank (Figure 1.4), which was commissioned in 1957, was the largest in the world. At that time it cost under $2 million. This was surpassed in 1970 by the German 330-foot reflector near Bonn, built at a price of $10 million. In Great Britain in 1973 detailed engineering designs were completed for an even larger telescope, 375 feet in diameter, for work in conjunction with Jodrell Bank. The lowest tenders for

FIGURE 1.4. *The 250-foot-aperture steerable radio telescope at Jodrell Bank. Commissioned in 1957, this was the largest completely steerable instrument in the world until the 330-foot reflector at Effelsburg, near Bonn, West Germany, came into use in 1970.*

the construction, received early in 1974, indicated that the cost of the facility would be about $60 million, and the project had to be abandoned.

The engineering difficulties are not only those associated with the large size of these telescopes and the requirement that they be steerable. Nowadays it is desirable to cover the entire band of radio wavelengths down to the millimeter region of the spectrum, where atmospheric absorption creates problems. The resolving power of a telescope of given aperture improves as the wavelength is decreased; furthermore, at these short wavelengths the spectral lines from molecular constituents in the gas clouds in the Milky Way can be studied. However, the shorter the wavelength, the greater mechanical accuracy is required in the paraboloidal surface. The specifications for the abandoned British telescope called for a deviation from the true paraboloid of not more than 2 millimeters over the entire surface—to be maintained at all elevations and in wind speeds of 20 miles per hour.

Although it is not possible to equal the adaptability of the large steerable reflector for radio astronomical research, the engineering problems and high cost have stimulated the development of other means for the achievement of resolving power. Very early in the developments after World War II, the principle of the radio interferometer was used. In Cambridge, Ryle and his colleagues placed two simple aerials a distance apart, connected by cable to a common receiver. This radio equivalent of the Michelson optical interferometer receives in a lobed polar diagram instead of a broadbeam; the angular width of the lobes is proportional to the wavelength and inversely proportional to the distance apart of the aerials. For example, if two aerials are separated by 3.5 kilometers and the wavelength is 1 meter, then the lobe width is about 1 minute of arc. If the radio waves received emanate from an extensive region, then as the rotation of the Earth sweeps the lobe pattern over the sky, the signal received will remain steady. On the other hand if the source has an angular diameter smaller than, or comparable with, the lobe width, the maxima and minima will be observed in the received signal strength as the Earth rotates. By varying the spacing it is possible to obtain information about the diameter of the emitting source. An equivalent principle was used simultaneously by the Australians, although in their case one aerial was mounted on a cliff site near Sydney, and the sea was used as a reflector. This is the equivalent of the optical Lloyds mirror interferometer; the baseline is equal to twice the height of the aerial above the sea.

By using interferometric systems it is possible to realize high resolving powers simply by moving the aerials farther apart. A major difficulty is that phase stability has to be maintained across the system. As long as the aerials can be connected by cable this is not a serious problem. Eventually microwave radio links must be used, and systems using one or more such links were developed to combine aerials 100 miles apart as an interferometer. Then, in 1967, the use of atomic clocks as reference enabled the signals at the separate aerials to be recorded on tape and combined subsequently, without any connection between the aerials. Success was achieved first by the Canadians. In 1969 American and Soviet radio astronomers combined their aerials in this way to record independently between the Soviet Union and America, and resolving powers of 0.0004 seconds of arc were realized.

In principle the aerial systems can be simple. In practice large aerials are required to obtain the necessary sensitivity, and large steerable paraboloids are commonly employed. If the source is tracked across the sky it is possible to derive information about its structure by complex analytical procedures. Many other derivatives of straightforward dish-type and interferometric systems have been developed. Among these the concept of aperture synthesis pioneered by Ryle in Cambridge has been most important. The principle of this system is that a large aperture is simulated by successive recordings of the signals received in two or more aerials, which are moved to a number of points within the simulated aperture. With high phase stability and computer processing, regions of the sky can be observed with an accuracy equivalent to an aperture of many kilometers. The most advanced Cambridge system uses a number of paraboloid aerials on a railway track 5 kilometers long (Figure 1.5). These aperture-synthesis aerials are not cheap—the Cambridge

FIGURE 1.5. *The 5-kilometer aperture-synthesis radio telescope at Cambridge, England, is a system of eight 13-meter steerable telescopes. Four of the dishes visible in the photograph move on a railway track $\frac{3}{4}$ mile in length. (Reproduced by permission of British Insulated Callender's Cables Ltd and the Director, Mullard Radio Astronomy Laboratories, Cavendish Laboratory, Cambridge.)*

5-kilometer system cost over $6 million in 1972. The Dutch have built an equivalent system at Westerbork in Holland. Perhaps the ultimate form of this system is now under construction in New Mexico for the National Radio Observatory. This is known as the Very Large Array (VLA); it is to consist of 27 steerable paraboloids, each 82 feet in diameter, moving on railway tracks in the form of three arms, each 21 kilometers in length, in the shape of a Y. By observing for 12 hours at a time as the Earth rotates, it is planned to synthesize radio maps with a resolution of 1 second of arc at a wavelength of 11 centimeters. The present (1974) estimated cost is $77.1 million.

Modern developments in radio astronomy in many countries now involve multi-million-dollar expenditures, with complex computer programming of the observations and the results. However, it must be remembered that the most remarkable discovery since that by Jansky of the existence of the radio waves was made in 1967 with cheap equipment, by Hewish at Cambridge. With an aerial array that cost less than $50,000, using simple pen-and-ink recorders, he discovered pulsars. Also the important discovery in 1965 of the microwave background radiation was made at the Bell Telephone Laboratories with a 20-foot-aperture horn antenna. This system was built for radio communication via Earth satellites, and the discovery, which has been of such importance to cosmology, was an incidental byproduct of this work.

THE COST OF ASTRONOMICAL RESEARCH RELATIVE TO NATIONAL SCIENCE BUDGETS

Estimates of the probable cost of the Soviet 236-inch optical telescope facility are in the region of $400 million. There is naturally uncertainty attached to this figure in relation to expenditures elsewhere, because of the difficulty of relating ruble and dollar costs. Further, it is not clear what fraction of this estimate refers to the telescope and its essential facilities. Elsewhere, expenditures on individual ground-based astronomical facilities are an order of magnitude less. Major ground-based optical telescopes can still be built for capital expenditures of the order of $20 million. In fact, the most expensive ground-based astronomical instrument at present under construction is the VLA radio telescope in New Mexico, the provision for which is $77.1 million (1974).

Astronomical research in space vehicles has created a discontinuity in expenditure. The estimated cost of the 120-inch Large Space Telescope is about $500 million,* but this would ride on the Space Shuttle, so that the real cost could only be estimated if an appropriate fraction of the ultimate development and production cost of the Shuttle were included. The problem of including an appropriate proportion of the cost of development and production of the launching rockets creates a real difficulty in assessing the costs of astronomical research in space. The cost of the abandoned High Energy Astronomical Observatory was estimated (1973) at $275 million for each payload. However, the cost is far less for small astronomical satellites placed in Earth orbit. For example, the British UK5 (Ariel 5) X-ray satellite launched on 15 Ocotber 1974 cost about $12 million, although this was the cost of the payload only, since the launching was made by NASA. The UHURU satellite, which yielded the remarkable results on the galactic and extragalactic X-ray sources, cost $13 million (1970), including the launching rockets.

Although individual small Earth-satellite astronomical projects are not necessarily more expensive than a major ground-based optical telescope, their scheduled lifetime is, at present, measured in years rather than decades. Hence the finance of astronomy in the United States in recent years has been dominated by the astronomical space activities of NASA. The 1972 report of the National Academy of Sciences (the Greenstein report) contains some revealing figures.† For example, 15 major radio-astronomical projects initiated or under construction in the period from 1958 to 1970

---

*See also footnote, p. 11.

†*Astronomy and astrophysics for the 1970's,* National Academy of Sciences, Washington, D.C., 1972.

TABLE 1 *Federal support for astronomy in fiscal 1972 (in $ millions)*

	Total	*Ground-based*	*Space*
National Aeronautics and Space Administration (NASA)	110.0	9.0	101.0
National Science Foundation (NSF)	29.9	29.9	—
Department of Defense (DoD)	9.0	7.0	2.0
Smithsonian Institution	3.0	1.5	1.5
Total	151.9	47.4	104.5

cost a total of $53 million, and 12 major optical and solar telescopes in the same period cost $54 million.* Against this total capital cost of about $107 million for the major ground-based American astronomical facilities in the 12 years from 1958 to 1970, NASA launched 7 successful astronomical satellites in the 3 years from 1967 to 1970 at a cost of $198 million. However, these are the only identified "direct costs." The report quotes the NASA estimate that during the fiscal years 1969 to 1971 "NASA has obligated or will obligate between $10 million and $15 million . . . annually on astronomical work that is part of, and budgeted through, its Lunar and Planetary Program. It is also important to note that (during fiscal years 1969–1971), out of its general costs for vehicle tracking, data acquisition, and administrative costs (including the costs of operating its research centers), NASA estimates that between $90 million and $100 million annually should be apportioned as the 'indirect cost' of maintaining its astronomy programs."

From this same report the information in Table 1 can be derived regarding the total federal support for astronomy in the fiscal year 1972, and its division into space and ground-based support.

Over the years 1966 to 1972 the fraction of the total budget

*The total capital cost of the National Radio Astronomy Observatory since its origin 21 years ago has been given as $39 million, and the total operating cost as $48 million (J. W. Findlay, *Sky and Telescope* 48, 360, 1974).

devoted to ground-based astronomy, between 20 and 30 percent, has remained about the same. The total amount has also remained substantially constant although there have been fluctuations in the amounts from the individual agencies. This federal support of the order of $150 million per annum does not, of course, represent the total support for astronomy in the United States. The estimate of support from academic institutions is subject to considerable uncertainty but can be placed at about $70 million per annum, of which $30 million is from nonfederal sources. Thus the overall figure for the support of all forms of astronomical research in the United States, both ground-based and in space, is in the region of $180 million per annum, to which must be added a further $100 million per annum if the NASA overhead support costs are included.

These sums are a small fraction of the total science budget of the various agencies. During the fiscal year that is taken as an example, the total NASA budget was about $3,000 million. The research and development budgets amounted to about $2,500 million, of which one half was allocated to manned space flight. Space science at $553 million was about 22 percent of the research and development budget. Hence the NASA contribution to astronomy is only about 4 percent of its annual research and development budget.* The proportion of the total budget of the NSF spent on astronomy is not dissimilar. With an annual expenditure of the order of $600 million, the expenditure on astronomy of $29.9 million represents about 5 percent of the total.

The total research and development expenditure in the United States in 1972 was approximately $15,500 million (Defense, $8,000 million; NASA, $3,000 million; other, $4,500 million). Thus the estimate of $180 million spent by all agencies on astronomy in 1972 represents just over 1 percent of the total U.S. research and de-

---

*A detailed breakdown of the NASA research and development budget for the years 1971 to 1973 is given in Table 7 of *The Origins and International Economics of Space Exploration,* by B. Lovell, Edinburgh University Press and John Wiley & Sons Inc., New York, 1973.

velopment budget. On the national scale these figures are extremely small. Taking the gross national product of the United States as about $1,000 billion, the overall research and development budget is about 1.5 percent and the total budget for all forms of astronomy is about 0.02 percent of the GNP. (For comparison, the U.S. defense budget is about 8 percent of the GNP and the education budget about 6 percent.)

Although in years subsequent to 1972 the nominal figures for research and development have increased substantially, the above percentages remain approximately the same. For example, in the President's budget proposals for 1975 the figures given relevant to research and development are shown in Table 2. The figures for 1974 and 1975 are estimates.

Furthermore, according to the National Science Foundation, although the numerical expenditures have increased, the actual purchasing power has decreased. The NSF estimate is that the actual purchasing power in the total research and development budget has declined from about $17,500 million in 1967 to about $13,000 million in 1974.

TABLE 2 *U.S. federal research and development budgets (in $ millions)* *

	1973	*1974*	*1975*
Defense	8,417	8,676	9,201
NASA	3,271	3,104	3,173
NSF	428	460	538
Other †	4,668	5,345	6,104
Total	16,784	17,585	19,016

*These figures are taken from *Nature 247*, 328, 1974.

†The major amounts under this heading go to Health, Education, and Welfare ($2,191 million in 1974) and the Atomic Energy Commission ($1,429 million in 1974).

In the United Kingdom the 1973–1974 estimate for state-supported research and development was $2,340 million.* Taking the GNP as $125 billion, this represents about 1.9 percent of the GNP. The state-supported science budget was $320 million, of which $160 million was allocated to the Science Research Council, the body responsible for the support of fundamental researches in the natural sciences. The allocation to space research and astronomy was $35 million (including $8 million to the European Space Research Organization). Thus the U.K. expenditure on all forms of astronomy was approximately 1.5 percent of the research and development budget, or about 0.02 percent of the GNP. (For comparison the U.K. defense and education budgets are each about 6 percent of the GNP.)

Thus, when considered as a percentage of the national incomes the expenditures in the United States and the United Kingdom are remarkably similar, namely an overall state-supported research and development budget amounting to 1.5 to 2.0 percent of the gross national product. The support of all forms of astronomical research, both ground-based and in space, amounts to 0.02 to 0.03 percent of the GNP.

*Great Britain, House of Commons, *First report of the Advisory Board for the Research Councils,* Cmnd. 5633, H.M.S.O., London, 1974; *Science Research Council report for the year 1972–1973,* H.M.S.O., London, 1973; *Annual abstract of statistics,* H.M.S.O., London, 1973. A conversion factor of £1 = $2.5 has been used.

CHAPTER TWO

THE SOLAR SYSTEM

In 1973 astronomers celebrated the 500th anniversary of the birth of Copernicus. He was not the first to have the idea that the Earth was in motion around the Sun. Aristarchus of Samos had made a similar suggestion one and a half thousand years earlier. However, the idea of Aristarchus did not survive and it was Copernicus who, in his famous *De revolutionibus orbium caelestium*, published in the year of his death in 1543, began the line of thought and development that finally destroyed the Ptolemaic doctrine that the Earth was fixed at the center of the universe.

Ptolemy had, indeed, provided a surprisingly good model, with the epicycles and deferents, for the explanation of the motion of the planets across the sky. The Copernican model, with the Earth and planets moving in circular orbits around the Sun, did not account as satisfactorily for the observed motions. However, this heliocentric doctrine was promulgated at a time of opportunity. Tycho Brahe, born in 1546, three years after the death of Copernicus, made far more accurate measurements of the planetary motions than had ever been made before. Although Tycho Brahe proposed a system in which the Earth was fixed, his life overlapped those of Galileo and Kepler. The observational evidence obtained by Galileo early in the 17th century for the motion of the Earth, and the realization by Kepler, who had access to Tycho Brahe's accurate measurements, that the Earth and the planets moved in ellipses, not circles, around the Sun, finally established the heliocentric concept.

It might be thought that after such a prolonged preoccupation with the motion of the bodies of the solar system man's interest would turn to their physical nature. However, this was not the case. The techniques of observation for physical investigations, notably the spectroscope, still had to be developed; it was 1814 before Fraunhofer discovered the spectral lines, named after him, in the light from the Sun, and thereby began the development of astrophysical spectroscopy.

The Ptolemaic and the Copernican systems were based on a system containing the planets Jupiter, Saturn, Mars, Venus, and Mercury as well as the Sun, Earth, and Moon. Kepler enunciated his laws of planetary motion in 1609, and Newton showed in his *Principia mathematica* in 1687 that they followed from the universal laws of gravitation. Nearly a century elapsed before Herschel, in 1781, discovered the planet Uranus, lying beyond the orbit of Saturn. Then in 1846 the planet Neptune was discovered, on the basis of calculations made by J. C. Adams of Cambridge, in 1845, and Le Verrier in Paris, in 1846, that the irregularities in the motion of Uranus were caused by another, undiscovered, planet. Pluto the, most distant of the planets, was discovered in 1930.*

Interest in the dynamical aspects of the solar system continues

---

*A body of contemporary opinion believes that Pluto is an escaped satellite of Neptune.

to this day. Not all the planetary motions are precisely described by the general theory of relativity, which in 1919 elucidated the problem of the motion of the perihelion of the planet Mercury, and there is recurrent speculation about the existence of a further planet, beyond Pluto.* When the technical possibility arose, just over a decade ago, of sending space probes to the planets Venus and Mars, the distance of these planets was not known with sufficient accuracy. A calculation of this depends on a precise knowledge of the solar parallax—the angle subtended by the radius of the Earth at the Sun. If this is known, the Keplerian laws can then be used to predict the precise distance of a planet at any moment. Attempts at a precise measurement of this parallax by various means had been made since Halley's first observations of the transit of Venus across the Sun's disc in 1716. Until 1961 the most modern measurements of the parallax differed among themselves by 0.1 percent, but in that year three separate radar measurements of the distance to the planet Venus established a definitive value, and the value of the astronomical unit (the mean distance from the Earth to the Sun) was subsequently derived to an accuracy of 1 part in 30 million.

Although there is a continuing interest in the purely dynamical aspects of the solar system, our age is remarkable for the development of a great interest in the physical condition of the solar system and the problem of its origin. The stimulus has been new techniques of observation, especially the radio telescope and the space probe. The case of the solar parallax has been mentioned. The same radar measurements simultaneously settled the problem of the rotation of the planet Venus. Since the planet is completely cloud-covered, no visual or photographic techniques could determine this rotation rate, and estimates varied from 24 hours to 225 days. Radar waves have penetrated the clouds, and the frequency spread introduced by the reflection from the rotating planet enabled an accurate

*A recent discussion of the small differences between the predicted and measured positions of Neptune since its discovery in 1846, and the possible implications as regards the existence of a 10th planet, has been given by D. Rawlins and M. Hammerton, *Monthly Notices of the Royal Astronomical Society 162,* 261, 1973.

determination to be made of the rotation. The sidereal rotation period has been found to be 244 days; moreover, the rotation is retrograde.

The enigma of the rotation rate of the planet was paralleled by speculation about the nature of the planetary atmosphere and surface conditions. Fifty years ago it was thought that the clouds were of water, like the clouds on Earth, and that the planet might be an abode of luxuriant vegetation and other forms of life. In 1932 spectroscopic observations from Mt. Wilson indicated that carbon dioxide might be present in the atmosphere of the planet. The difficulty was that all such attempts to study the spectrum of the planet had to be made through the Earth's atmosphere. After World War II, attempts were made to obtain more decisive information about the existence of water or oxygen in the Venusian atmosphere by measurements made from high-flying balloons. Eventually radio telescopes measured the short-wave radio emission from the planet, and this indicated that the surface temperature might be very high —in the region of 470 to 570 K. However, knowledge about the constitution of the atmosphere and the rotation rate was so meager that extreme speculations were made about the planetary surface. As recently as 1955 distinguished astronomers were publishing scientific papers in support of the view that the surface of the planet was completely covered by a boiling ocean, or alternatively, that it was entirely an arid desert.

These extreme conjectures were settled by the Soviet space probe to the planet, Venera 4. Launched on 12 June 1967, the probe released a capsule as it entered the planetary atmosphere at 0734 Moscow time on 18 October 1967. A parachute system operated to slow the descent of this capsule through the atmosphere. As it descended to the planetary surface signals were transmitted to Earth for 96 minutes, giving information about the constitution, pressure, and temperature of the atmosphere. Originally it was assumed that the signals ceased when the capsule landed on the surface, but subsequent analysis indicated that the device probably ceased operating when it was still about 24 kilometers above the surface. The temperature at that point was very high (536 to 551 K), the pressure was 20 times that of the atmospheric pressure on Earth, and the atmosphere contained over 90 percent of carbon dioxide. The American Mariner 5 spacecraft passed the planet a day later, and similar measurements of high pressure and temperature and a large quantity of carbon dioxide were made from the remote sensors in that spacecraft.

The Russians again sent probes to Venus in 1969—Venera 5, launched on 5 January, and Venera 6, on 10 January. These were designed to release similar capsules to descend through the planetary atmosphere, which they did on 16 and 17 May respectively. From these measurements it was concluded that the atmosphere was 97 percent carbon dioxide, that the pressure at the surface was 100 atmospheres, and that the temperature was 773 K. At the next launch window, Venera 7 was dispatched to the planet on 17 August 1970, and descended through the atmosphere on 15 December. Although telemetry apparently ceased after 35 minutes, the Russians announced on 26 January 1971 that a computer study of the signals subsequently received indicated the reception of weak telemetry for a further 23 minutes (at 1 percent of the signal strength during the descent phase). Whatever the fate of the previous capsules, there seems no doubt that Venera 7 definitely landed and transmitted from the Venusian surface, confirming the previous high values of temperature and pressure.

The most recent in this series of Soviet probes to Venus (Venera 8), launched on 27 March 1972, released a capsule that descended through the planetary atmosphere on 22 July 1972 and landed on the surface of Venus near the equator, 500 kilometers from the dawn terminator. In addition to repeating previous measurements of the atmospheric pressure and temperature, Venera 8 transmitted data that enabled the winds to be calculated.* At the beginning of the parachute descent (54 kilometers above the surface) winds of 130 to 140 meters per second (about 30 to 40 miles per hour) were measured. The wind decreased with decreasing altitude and was

*M. Ya. Marov et al., *Doklady Akademiia Nauk S.S.S.R. 210,* 559, 1973; summarized in *Nature, Physical Science 244,* 17, 1973.

only 6.4 meters per second (14.5 miles per hour) at the surface. The temperature at the point of impact was 741±7 K, the same as recorded by Venera 7 on the dark side (747±20 K), suggesting that there is little diurnal temperature variation, no strong thermal wind across the terminator, and hence little dust or surface erosion. During the descent the altitude calculated from hydrostatic information differed from the values given by the radar altimeter, and one possibility is that during the descent the capsule was drifting over an incline of about 7 degrees.*

This direct revelation of the hostile conditions on the planet Venus is in sharp contrast to the speculations of a few decades ago that the cloud-covered planet was a favorable abode for vegetation, and perhaps other forms of life. The entirely different evolutionary paths of Earth and Venus are a fascinating problem for contemporary studies of the evolution of the solar system. Both bodies have almost identical physical characteristics. The diameter of Venus is only 3 percent less than that of the Earth. Its mass is 18.5 percent less; the mean density (4.95 gm cm^{-3}) is not dissimilar from that of the Earth (5.52 gm cm^{-3}); the surface gravity (850 cm s^{-2}) and escape velocity (10.3 km s^{-1}) are similar to those of the Earth (982 cm s^{-2} and 11.2 km s^{-1}). These differences are minor compared with the differences between Earth or Venus and the other planets (the nearest equivalent being Mars, which, compared with Earth, has radius 47 percent less, mass 90 percent less, mean density 3.95 gm cm^{-3}, surface gravity 376 cm s^{-2} and escape velocity 5.0 km s^{-1}). Why then have Earth and Venus, with such similar characteristics, evolved along entirely different paths? The one is a generally pleasant abode for life, the other hostile in the extreme, with a

*No U.S. spacecraft has yet attempted to land instruments on the surface of Venus. Of the attempted flights to the vicinity of the planet, Mariner 2 (launched 27 August 1962) passed the planet at a distance of 34,700 kilometers and Mariner 5 (launched 14 June 1967) at a distance of 4,000 kilometers. Both of these probes returned valuable data about the environment of the planet and the atmosphere, obtained by remote sensing. An earlier Venus flight attempted by the United States (Mariner 1, launched 22 July 1962) failed. The relative success of the U.S. and the U.S.S.R. in studying the planets has been reversed in the case of Mars.

poisonous atmosphere of carbon dioxide at 100 times the atmospheric pressure at the surface of the Earth, and a surface temperature so high that lead would be molten and mercury would long since have boiled.

No one knows the answer to this question. The current view is that the differences arise because Venus moves in an orbit that is somewhat closer to the Sun than the Earth's.* Although the bodies may have been similar in the beginning, probably the higher temperature of Venus facilitated production of carbon dioxide, which in turn began to produce a greenhouse effect in the atmosphere; hence the situation would have worsened rapidly. Whether this is, in fact, the explanation, and if so at what stage in the evolutionary time sequence the significant differences began to occur, is an intriguing problem, which may be solved as more detailed information is accumulated by future space probes to the planet. One would especially like to know if the significant divergence began after the initial stages of organic evolution on planet Venus had occurred.

In the case of the planet Mars, too, modern techniques have reopened problems of vital interest. Although this planet moves in an orbit more distant from the Sun than the Earth's,† surface features have been distinguishable within the limits of resolution of the terrestrial telescopes. Further, because of the eccentricity of the orbit, occasional oppositions bring the planet close to Earth—in September 1956, for example, the distance between Earth and Mars was only 56 million kilometers. During the oppositions of 1877 and 1888 Schiaparelli observed many long straight intersecting lines, changing in size with time. The American astronomer Percival Lowell made a speciality of studying these "canals" from the Flagstaff Observatory and charted more than 400 of them, some over 4800 kilometers in length. A widespread view formed that these were irrigation systems, associated with the melting of the polar caps of

*The semi-major axis of the Venus orbit is 108.21 x 10^6 kilometers, compared with 149.60 x 10^6 kilometers for the Earth.

†The semi-major axis of the Martian orbit is 227.9 x 10^6 kilometers, compared with 149.60 x 10^6 kilometers for the Earth.

the planet. At the 1956 opposition Sinton made spectroscopic investigations using the 200-inch Hale telescope, and these supported the idea that vegetation exists on some regions of the planet. The noon surface temperatures at the equator are computed to range from 268 K to 298 K, and although nitrogen is believed to be a major constituent of the atmosphere, there is no irretrievably poisonous condition of noxious gas, pressure, or temperature, such as exists on Venus. It is understandable, therefore, that Mars has been a prime target of study in the search for extraterrestrial life forms. Until the advent of the space probe, the great difficulty has been that the best resolution given by terrestrial telescopes has only been of the order of 50 to 70 kilometers, even under the best conditions.*

As soon as space techniques had been sufficiently developed, it was natural that both the United States and the U.S.S.R. should attempt to send probes to Mars. In contrast to the space probes to Venus it is the American probes that have scored notable successes, while the Soviet probes have failed in their primary missions. The first successful probe to Mars was the American Mariner 4, launched on 28 November 1964, which passed the planet at a distance of 9,800 kilometers on 15 July 1965, and transmitted to Earth historic photographs revealing the surprising fact that the surface of the planet was heavily cratered. Mariners 6 and 7, launched on 24 February and 27 March 1969, were similarly successful, passing the planet at ranges of 3,200 kilometers and 3,490 kilometers on 31 July and 5 August respectively. However, by far the most remarkable is the latest in the series, Mariner 9, which was placed in orbit around the planet on 13 November 1971.† In spite of a prolonged dust storm that obscured the features of the planet for some time after the orbit was achieved, this probe eventually transmitted to Earth high-quality pictures of Mars and its moons, together with other data on the properties of the planet. The probe was placed in an orbit around the planet with a period of 11 hours 58 minutes,

---

*At the favorable 1956 opposition the planet appeared to terrestrial observers as a disc 25 seconds of arc in diameter—equivalent to viewing a cent at a distance of 135 yards.

†Mariner 8, launched on 8 May 1971 and destined for Mars, failed.

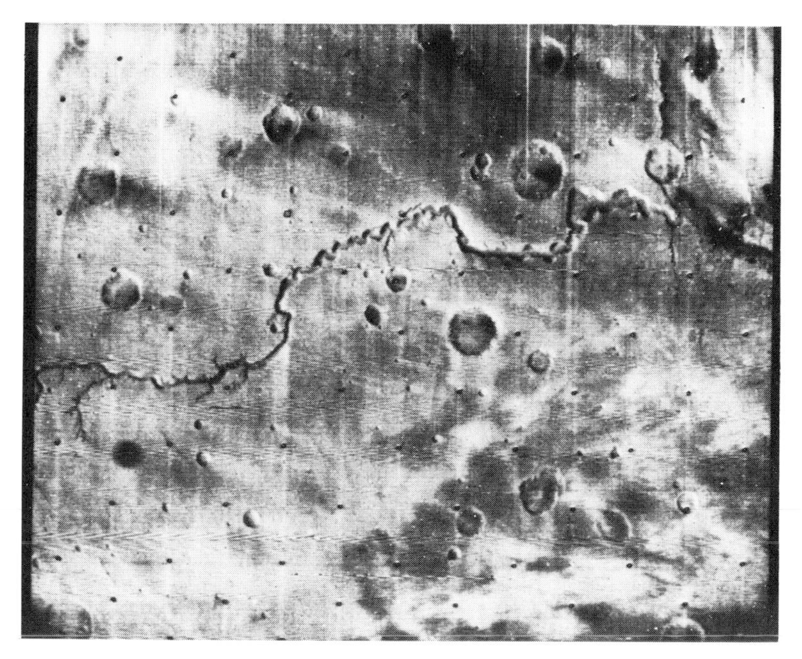

FIGURE 2.1. *A Mariner 9 photograph of the Martian surface. The sinuous valley in the Mare Erythraeum is about 250 miles long and 3 miles wide. It resembles a terrestrial gully cut in the ground by running water, and may be a dried river valley. (JPL/NASA)*

1,400 kilometers from the planet at the lowest part of the orbit and 17,000 kilometers at the most distant. Pictures were transmitted until 27 October 1972, by which time 7,329 pictures had been received on Earth, obtained during 698 circuits of the planet.* Complete photographic coverage was secured using wide-angle lenses with a resolution of 1 kilometer, and many regions have been covered in greater detail with narrow-angle lenses giving a resolution of 100 meters, or 500 times better than the best of the photographs hitherto taken from Earth by telescopes (Figures 2.1, 2.2).

*Although the spacecraft's transmitter was turned off by ground control from Earth on 27 October when the on-board supply of nitrogen gas ran out, Mariner is expected to continue in orbit around the planet for another 50 to 100 years.

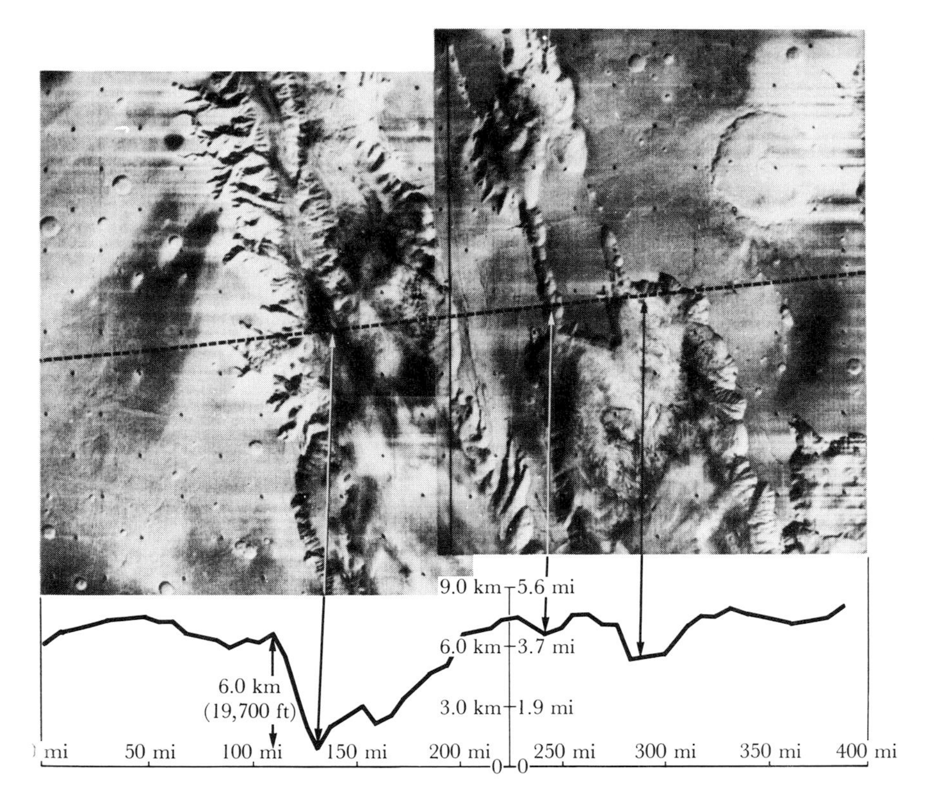

FIGURE 2.2. *A part of the Tithonius Lacus chasm on Mars, photographed by Mariner 9 from an altitude of 1070 miles. The ultraviolet spectrometer in Mariner 9 scanned along the dotted line, measuring the atmospheric pressure at the surface of the planet. The relief outline below the photograph shows the transformation of these measurements to relative depths. The lowest point, 19,700 feet below the adjacent terrain, is nearly four times the depth of the Grand Canyon in Arizona (5,500 feet). (JPL/NASA)*

Clearly, much time will elapse before a detailed interpretation of this great mass of data can be made, although the Mariner 9 coverage has already led to a revision of the idea, engendered by the photographs from the previous Mariners, that the planet is heavily cratered overall. It is now apparent that these earlier photographs were, by chance, of the heavily cratered regions, and that at least two geological provinces must be distinguished—the heavily cratered regions and the smooth regions, which are coincident with the classical "deserts" in many cases. Some of the mountain peaks rise to 26,000 feet, and there are enormous cratered terrains, the Hellas basin, for example, having twice the area of Texas and a floor 2 or 3 kilometers below the level of the surrounding terrain.

Notwithstanding these revolutionary advances in our detailed knowledge of the planet, there remain large areas of uncertainty. One vital question is that of the existence of water on the planet. The earlier Mariner photographs revealed a cratered surface so similar to lunar features that the view that any water existed in the polar caps or elsewhere received a severe setback. But the Mariner 9 photographs showed that, by coincidence, these earlier photographs were not representative of the whole surface, and the higher-definition photographs have revealed features that definitely seem to indicate the existence of water, at least at some stage in the history of the planet. Undoubtedly one of the most exciting revelations of the Mariner 9 exercise was the existence, near the canyons, of some features so closely resembling river beds that, contrary to earlier views, it is now believed that much water may be locked up in the polar caps or in subsurface permafrost.

This new evidence for the presence of water on the planet lends a new significance to the proposed Viking 1975–1976 probes that NASA intends to dispatch to Mars, which will attempt to land working equipment including apparatus for biological investigations. The results of these landings will be awaited with particular interest, since they may lead to definite conclusions about the possibilities of organic development on Mars.

The present uncertainties about Mars are well illustrated by the protracted discussions that occurred for over a year within the team responsible for deciding where the landings should be made in the summer of 1976. Eventually four sites were chosen—a prime site and a back-up site for each lander. In a letter to *Nature*, Carl Sagan, a member of the selection team, emphasized the difficulty of choosing the best landing sites on the basis of photographs with

100-meter definition.* He drew an interesting comparison with the situation in reverse, that is, the choices being made for a hypothetical spacecraft from somewhere else to land on Earth. The coordinates of the four sites chosen on Mars would correspond on Earth to three ocean landings and one near the Yellowstone River. The three spacecraft landing in the ocean would probably sink and not even be noticed by the inhabitants of Earth. The fourth would provide geological data untypical of the world in general, but might reveal the presence of indigenous and perhaps intelligent life on Earth. In fact, given photographs of Earth with the same definition as those we now have of Mars these choices would be highly probable. With a 100-meter definition, Yellowstone might be chosen because of its probably high interest to geologists, while the oceans would be chosen because they would appear smooth at this resolution and would be regarded as safe landing sites. The conclusion is that many more than two Viking landers will be required before it becomes possible for men to characterize the principal features of the planet Mars.

The Viking landers in 1976 will not be the first probes to reach the planet. After many failures with Martian space probes, the Russians launched Mars 2 and Mars 3 on 19 and 28 May 1971 respectively.† Mars 2 reached the vicinity of the planet on 17 November, and Mars 3 on 2 December. A capsule was landed on the planet, but unfortunately the transmissions ceased almost immediately. Further attempts were made in 1973–1974. Mars 4 and 5 were launched on 21 July and 25 July 1973 respectively. There were expectations that an ambitious landing on the planet was planned. In the event, problems seem to have developed. Mars 4

---

*C. Sagan, *Nature 244,* 61, 1973.

†*Soviet space programs 1966–70,* U. S. Senate Document 92-51, Washington, D.C., 9 December 1971, lists on p. 219 four unannounced launchings (10 and 14 October 1960, 24 October and 4 November 1962) believed to be Mars probes, which either failed to reach or failed to leave Earth orbit. Mars 1 (1 November 1960) passed Mars at a distance of 193,000 kilometers, but the communications failed; Zond 2 (30 November 1964) passed within 1,500 kilometers of the planet, but again the communications failed. The list also includes Zond 3 (18 July 1965), launched towards Mars as an engineering test. This probe returned photographs of the far side of the Moon.

could not be placed in orbit, and it flew by at a distance of 2,200 kilometers on 10 February 1974. Mars 5 entered a Mars orbit on 12 February 1974. The main information from this series of Soviet Mars probes has not been from landers on the surface as expected, but from remote sensors and photography of the planet by the orbiters. An account of the results from Mars 2 and 3, and from the Mariners 6, 7, and 9, on the Martian atmosphere has been published by Soviet scientists.*

The American mastery of planetary investigation by unmanned space probes has been demonstrated further in the launching of Pioneer 10 and Mariner 10. On 3 March 1972 Pioneer 10 was launched towards the planet Jupiter. After a flight of 21 months, during which it passed through the asteroid belt, Pioneer 10 made a fly-by of Jupiter on 4 December 1973 at a distance of 130,400 kilometers. The orbit was calculated so that, in the momentum exchange with Jupiter, Pioneer 10 was accelerated and is now on its way out of the solar system.† Throughout its journey Pioneer 10 transmitted valuable information about conditions in interplanetary space, but of course the major interest pertained to its close approach to Jupiter, and indeed to whether the instrument would survive the intense radiation field of the planet. It did so, and the information telemetered to Earth during this brief encounter is changing many views about the planet. The radiometer in Pioneer 10 showed that the planet is radiating two or three times as much energy as it receives from the Sun. One view is that this additional energy is derived from the continuing gravitational contraction of the planet. An entirely new discovery was that the dark orange brown bands are troughs within which cooler atmospheric gases are descending. The Great Red Spot may be an intense updraft of gas analogous to

*K. Ya. Kandrat'yev and A. M. Bunakova, *Meteorologiya Marsa,* Hydrometeorological Press, Leningrad, 1973, translated in NASA Document TT F-816, Washington, D.C., 1974.

†On the remote chance that at some future time another civilization in the galaxy will encounter Pioneer 10, a plaque containing information about our position in the galaxy was fixed on the antenna support struts of the probe. The details have been given by C. Sagan, L. S. Sagan, and F. Drake, *Science 175*, 881, 1972.

a terrestrial hurricane (Figure 2.3). The results have also stimulated new arguments about the internal structure of the planet. Some planetary scientists believe that Jupiter has a solid rocky core of 40 Earth masses, surrounded by a thick mantle of ice and metallic hydrogen. Another view is that there is no solid core but that below about 25,000 kilometers the planet is mainly metallic hydrogen.

More data about the planet are expected from the fly-by of Pioneer 11 in December 1974. This probe, launched on 6 April 1973, was originally intended to follow Pioneer 10 out of the solar

FIGURE 2.3. *The planet Jupiter photographed by Pioneer 10 on 1 December 1973 when it was 1.6 million miles from the planet. The terminator is at the bottom. The Great Red Spot and the shadow of the satellite Io can be seen. (NASA)*

system after the Jupiter encounter. However in a revision of the plan the thrusters were fired in the spring of 1974, so that the probe would approach to within 42,000 kilometers of Jupiter's cloud tops; after encounter, the flight path should take it to the vicinity of Saturn in September 1979, that is, 6½ years after launching. Although this is well beyond the designed lifetime for the probe, it is hoped that sufficient equipment will still be operational to provide information about the nature of the rings and other features of the planet from this first close approach of scientific instruments.*

Mariner 10, launched on 3 November 1973, passed Venus at a distance of 5,750 kilometers on 5 February 1974, and transmitted many pictures of the cloud-covered planet. On 29 March 1974 the probe photographed Mercury from close range. The results have been dramatic. Hitherto the best Earthbound observations of the planet showed little but indistinct features, and the planet's correct rotation period of 59 days was not established until 1965, by radar measurements. The minimum distance of approach of the probe was 750 kilometers, on the dark side of the planet, and 2,300 photographs of the planet's surface were transmitted. From 35 minutes before close approach, to 22 minutes after, 36 pictures were obtained, with a resolution of 100 to 500 meters. The photographs revealed numerous craters, scarps, ridges, circular basins, and plains, in many cases with a strong resemblance to the lunar surface (Figure 2.4). The probe made a second close encounter with the planet on 21 September 1974, at a distance of 47,900 kilometers, during which a further 500 photographs were obtained. This extended the photographic coverage to about 37 percent of the surface. A third encounter of Mariner 10 with Mercury in March 1975 may further extend this coverage.

Even the preliminary studies indicate that Mercury has had a history similar to the Moon's, having suffered a period of heavy

---

*At 9:42 P.M. Pacific Standard Time on 2 December 1974 Pioneer 11 made its close approach to Jupiter—26,000 miles above the cloud tops. The encounter with Saturn is expected to occur on 2 or 3 September 1979 at a distance of 1.05 to 1.15 Saturn radii. After this encounter Pioneer 11 will leave the solar system in the direction opposite to Pioneer 10.

FIGURE 2.4. *The South Pole of the planet Mercury photographed by Mariner 10 from a distance of 53,200 miles, less than two hours after its second close encounter, on 21 September 1974. The pole is located inside the large crater (110 miles across) on the planet's limb (lower center). The crater floor is shadowed and its far rim, illuminated by the Sun, appears to be disconnected from the edge of the planet. (NASA)*

bombardment early in its history, followed by widespread volcanism. Mariner 10 carried other scientific equipment in addition to the cameras. Refined values for the mass and diameter of Mercury have been obtained, corresponding to a mean density of 5.44 gm cm^{-3}. It has a very tenuous atmosphere, probably mainly of neutral helium. The temperature measurements (100 K on the night side) and the manner in which the temperature declined after sunset on the planet indicate that the surface is covered with porous soil, like the Moon. The interpretation of the magnetic field measurements is not yet clear; the unexpectedly high value measured may be associated with a small intrinsic field of the planet, but it is probably largely produced in a complex manner by interaction with the solar wind.*

It is a remarkable feature of recent years that, in addition to the manned and unmanned excursions to the Moon, cameras and scientific instruments launched from Earth have closely inspected the planets Venus, Mars, Mercury, and Jupiter. There has resulted a great and revolutionary extension of the knowledge obtained by Earthbound telescopes. The full impact of the new techniques on our understanding of the evolution of the solar system cannot yet be assessed.

THE ORIGIN OF THE SOLAR SYSTEM

The Sun has long been recognized as a typical main-sequence star in the Milky Way. The processes by which a star like the Sun evolves from the primeval gas will be discussed later (Chapter 3). Here we are primarily concerned with the manner in which the Sun acquired its family of planets and whether the Sun is unique in this respect among the immense number of stars in the Milky Way and in other galaxies. It may be said at once that there is no complete understanding on this issue; indeed, over the past centuries many different theories have been proposed. In the late 18th and early 19th century the famous French mathematician Laplace proposed his nebular hypothesis. He believed that originally all the material that now forms the Sun and planets was a large nebula of rarefied gas in slow rotation. As the nebula cooled the gas contracted because of internal gravitational forces, and the rotation quickened. Eventually, the rotation became so rapid that some of the gas detached itself from the periphery, forming a ring outside the nebula. This happened repeatedly until eventually the

*A preliminary interpretation of the results of the Mariner 10 first encounter with Mercury was given in eight articles by the investigation team in *Science,* 12 July 1974.

central portion condensed into the Sun and the material in each ring formed a separate planet.

The theory of Laplace accounted for many of the general features of the solar system and it enjoyed a long period of success. However, Clerk Maxwell proved that such rings would not coalesce into planets, but would be transformed into a collection of much smaller bodies, like Saturn's rings. A second objection, which was to prove fatal to many subsequent theories, concerned the distribution of angular momentum in the solar system. The problem is that 99.9 percent of the mass of the solar system is in the Sun and it would be expected that nearly all the rotational momentum of the solar system would be carried by the Sun. On the contrary we find that the planets are moving very fast in their orbits compared with the rate of rotation of the Sun and that 98 percent of the momentum is carried by the planets. This segregation of mass and momentum must have occurred, on the Laplacian hypothesis, when the peripheral rings separated from the contracting nebula. If at that stage 98 percent of the momentum was carried in the rings, but only 0.1 percent of the mass, then enormous velocities must have been involved, and condensation even into small asteroids would have been impossible.

These searching criticisms of the nebular hypothesis in the second half of the 19th century led two American astronomers, Chamberlin and Moulton, to suggest quite a different theory of the origin of the solar system. Their hypothesis was that in the remote past the Sun was an ordinary star without planets, but that at some stage another star passed close to the Sun. The gravitational attraction between the Sun and this star swung them about one another, and eventually the other star passed on. In this encounter great amounts of gaseous matter would have been torn from the Sun, some of which would have remained under its gravitational influence, eventually condensing into small fragments, and finally accreting into larger bodies to form the planets. There were a number of variations of this hypothesis, such as the tidal theory of Jeans and Jeffreys. In this theory it was suggested that the passing star pulled out from the Sun a great filament of gas, which broke up and condensed straight away into the planets without the intervening accretion process suggested by Chamberlin and Moulton. In 1935 the American astronomer H. N. Russell undermined theories of this type when he showed that in any such encounter the star and Sun would have to approach one another so closely that the planets would eventually move in orbits many thousands of times closer to the Sun than is actually observed.

After the encounter theories were abandoned there was a partial return to the nebular hypothesis of Laplace. In its modern form the theory envisages a solar nebula consisting of a mixture of gas and dust. Whereas Laplace suggested that the planets condensed from a gaseous nebula, the current view is that the planets were formed by successive accretion through collision of the dust and gas of the nebula. As the primordial cloud rotated around the Sun the dust particles gradually became concentrated into a flattened disc. At each collision the relative velocity of the colliding particles would be decreased, the energy being radiated away as heat. Eventually, embryos formed in the cloud, and by successive collision, disintegration, and accretion the Earth and major planets evolved. A few billion years would be required for this process to mature. Today, collisions of the remaining dust still occur. Occasionally the larger bodies penetrate the atmosphere of the Earth and land as meteorites. Much more dust bombards the Earth, at the rate of millions of tons per year, but we are protected by the atmosphere, so that the dust either burns up as meteors, or is so small that it is stopped by the atmosphere before ablation occurs. The violence of the bombardment in the early history of the solar system is well evidenced by the cratering on the Moon, and on Mars and Mercury as revealed by the recent Mariner space-probe pictures (Figures 2.1, 2.2, and 2.4).

In the modern epoch there have been two distinct views about the origin of the solar nebula. One is that the Sun captured the nebular cloud when it passed through a dense region of interstellar dust. Otto Schmidt and the Russian school believed in this idea. Then Hoyle proposed a different capture idea—that the Sun was originally a member of a binary star and that this binary companion exploded. Some of the remnants of the disintegration would have been captured by the Sun to form the solar nebula. Probably the most generally agreed-upon modern view is that the solar nebula was not captured in any such manner, but that it is a natural consequence of the processes of star formation. Today it is believed that stars are not formed individually, but that they condense in large numbers from the primeval interstellar gas clouds according to the processes described in Chapter 3. When this occurs not all of the gas clouds will condense entirely into the stars. Some will remain as nebulae around the contracting stars, and from these nebulae planets form by the process of accretion described above.

A review of the many theories of the formation of the solar system proposed during this century has been published recently by Williams and Cremin.* Their conclusion is that the correct idea is probably a form of the theory described above, namely that the planets originated by accretion of the dust and gas in the solar nebula, and that the nebula itself arose as a natural consequence of the original formation of the Sun from an interstellar gas cloud. Of course, many difficulties remain. An acute problem is still that of the relative distribution of mass and angular momentum between the Sun and planets. One suggestion is that the solution to this lies in the effect of magnetic forces between the Sun and the planets during the evolutionary phases.

There is one dramatic consequence of the modern theories. On the abandoned encounter or near-collision theories that were popular fifty years ago, the solar system must have been almost unique in the universe. In fact, because a typical star is, say, a million miles in diameter, and separated from its nearest neighbor by over 20 trillion miles, it may be estimated that only one or two close encounters have occurred in the whole history of the galaxy. Thus, the formation of planetary systems by a close encounter of two stars would be an exceedingly rare occurrence, even when the whole universe is considered. On the other hand, the formation of the solar nebula

---

*I. P. Williams and A. W. Cremin, *Quarterly Journal of the Royal Astronomical Society* 9, 40, 1968.

as a natural consequence of the birth of the star that became our Sun implies that the majority of stars will have been formed with similar nebulae. Thus planetary systems must be expected to be a common feature of the stars in the Milky Way, and indeed in the entire universe.

Do we have evidence for the existence of planets around other stars in the Milky Way? As far as direct photography is concerned the answer must be negative. The largest Earthbound telescope has neither the sensitivity nor the resolution to detect a system of planets around even the nearest star to the Sun. There is, however, indirect evidence derived from a study of the motion of the nearby stars through space. For example, in 1963 P. Van de Kamp of Swarthmore College announced that after studying the motion of Barnard's star for 25 years he had concluded that the irregularities of its motion through space could be explained if the star was accompanied by a planet about the size of Jupiter. In 1969 he announced that a second, slightly smaller, planet must exist in orbit around Barnard's star. The star known as Lalande-21185 also shows an irregularity in its path across the heavens, indicative of the presence of a planet ten times as massive as Jupiter.

Evidence for planets in process of formation (protoplanets) around a star was published recently by astronomers from the Stockholm Observatory.* They made simultaneous photographs by three telescopes to study the light variation of the star RU Lupi. They found changes in the luminous output of 20 percent in one or two hours at all wavelengths, and also that the properties of the spectral lines were unchanged during these fluctuations in brightness. The stability of the spectral lines appears to rule out any intrinsic changes in RU Lupi itself, and it is suggested that the fluctuations are an extrinsic effect, caused by a swarm of protoplanets that periodically block out some of the light from the star.

---

*G. F. Gahm, H. L. Nordh, S. G. Olofsson, and N. C. J. Carlborg, *Astronomy and Astrophysics* 33, 399, 1974.

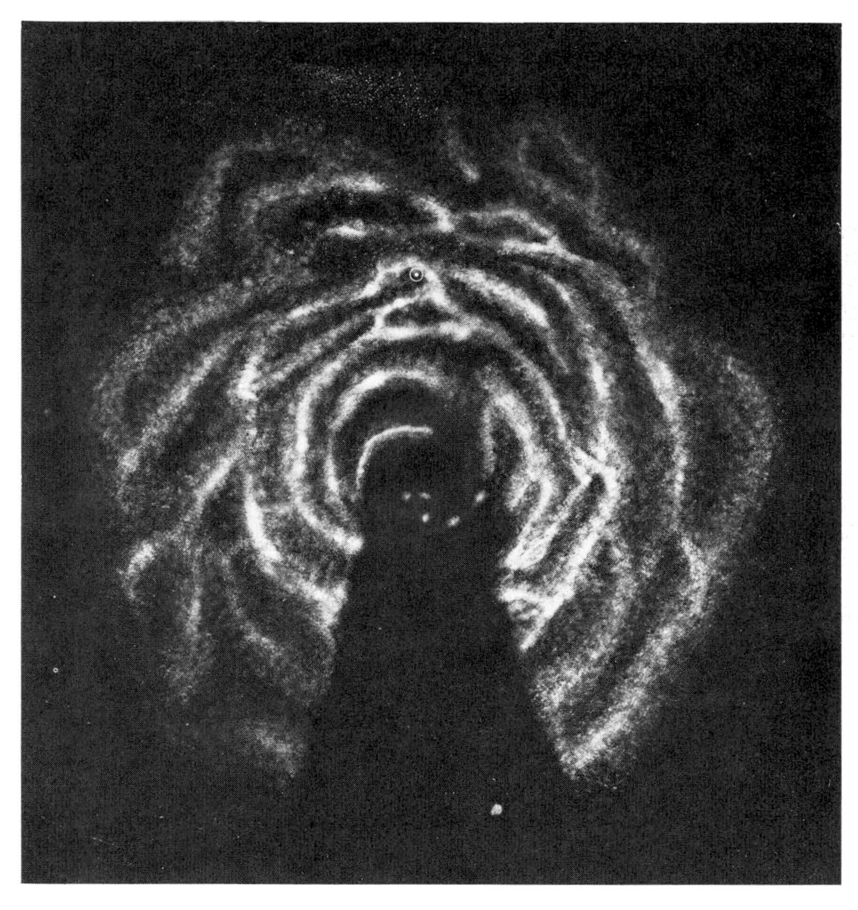

Artist's impression of the neutral hydrogen distribution in the galaxy, based on radio-telescope surveys of 21-centimeter wavelength radiation. The view is from the north galactic pole. The Sun (some 30,000 light years from the center) is marked by the symbol ⊙. (Courtesy of G. Westerhout, University of Maryland.)

CHAPTER THREE

THE MILKY WAY

HISTORICALLY THERE HAVE BEEN two critical epochs in the attempt to understand the universe. The first occurred in the 16th to the 17th centuries. Until that time it had been believed that the Earth was fixed in space and that it was at the center of the universe. In the epoch of Copernicus, Galileo, and Kepler this concept was destroyed and the heliocentric view established. With the publication of Newton's *Principia mathematica* in 1687 the full physical synthesis appeared to be complete, since, through the inverse-square law of gravitation, the Copernican and Keplerian theories appeared as natural consequences of a fundamental law of nature.

Throughout the whole of this period, and for over a century thereafter, little definitive progress was made in the attempt to understand the larger-scale structure of the universe. Although the heliocentric theory had superseded the geocentric concept as far as the Earth, Sun, and planets were concerned, nevertheless the fundamentally egocentric idea remained firmly in man's mind. The solar system itself was believed to be at the center of the system of stars; and further, it was believed that this symmetrical distribution of stars comprised the totality of the universe. The second critical epoch in the attempt to understand the universe was the decade following 1918, when both of these beliefs were shown to be erroneous. This revolutionary progress followed the discovery of a means for measuring the distances of stars that are far removed from the solar system.

It is true that Bessel in 1838 made the first measurement of the distance of a star by observing the apparent change in position of the star 61 Cygni, when viewed against the background of faint stars, as the Earth was at opposite points in its orbit around the Sun. He found that the parallax was 0.30 seconds of arc, and by straightforward trigonometry he calculated the distance to be 11 light years. But by the beginning of the 20th century, even with the advent of photographic techniques, it had been possible to extend Bessel's trigonometric method only to about 100 light years, which enabled the distances of a few thousand stars to be determined.

The major advance that enabled astronomers to extend these stellar distance measurements followed from a discovery made by Miss Henrietta Leavitt of Harvard in 1908. Miss Leavitt studied the variation in intensity of certain types of stars in the Small Magellanic Cloud. These stars are known as Cepheid variables, a class named after the star δ Cephei, which changes its brightness by a magnitude from maximum to minimum in 5 days. She noted the curious fact that the brighter stars took longer for their light variation from maximum to minimum and, indeed, found that there was a definite relationship between the brightness of the star and its period of light variation. Fortunately there were Cepheid variables among the stars whose trigonometrical parallax could be measured, and Harlow Shapley, working with the 60-inch telescope at Mt. Wilson, calibrated Miss Leavitt's apparent-brightness/period relationship in terms of absolute magnitude. It was then possible, by measuring the period of a Cepheid variable, to find its absolute magnitude, and hence, by observing its apparent magnitude, to determine its distance by the inverse-square law.

Shapley then studied the Cepheid variables in the globular clusters, and by 1918 had determined the distance of 25 of the 100 known objects of this type. He found that they were all at great distances from the Sun—15,000 to 100,000 light years. Now the globular clusters are unevenly distributed over the sky. They are all in one half of the sky, and a third are located in a small region around the great star cloud in Sagittarius. Shapley made the imaginative assumption that these globular clusters defined the general outline of the Milky Way system of stars, and he concluded that the Sun was far away from the center of the system. (Shapley's figure was 50,000 light years; the figure adopted today is 33,000 light years.)

Within a few years of the publication of these results in 1918–1919, the major conclusion that the Sun was far removed from the central region of the Milky Way had been wholly accepted and the age-old egocentric concept of man's place in the universe had finally been eradicated. There are about 100,000 million stars in the Milky Way. These are arranged in a flattened disc, with a dense central hub of stars, and several spiral arms of stars, gas, and dust radiating from this central region. The Sun is located in one of these spiral arms, 33,000 light years from the galactic center. Altogether, the system extends for about 100,000 light years. The dense central core is about 20,000 light years thick, but in the region of the spiral arms the dimension perpendicular to the disc is only about 5,000 light years. The system of stars is therefore highly flattened; moreover it is rotating, with the arms trailing like a viscous fluid, so that the speed of rotation varies with the distance from the galactic center. At the Sun's distance the system rotates once in 230 million years, but nearer to the galactic center the velocity of rotation is greater. For example, at a distance of 3,300 light years from the center the system rotates once in 28 million years.

From these rotational data it is possible to calculate the total mass of our galaxy. This mass is equivalent to 2×10^{11} Suns ($M_\odot$).

Formerly there was much discussion about the distribution of this total mass between the material in the form of stars and that in the form of interstellar gas and dust. Until the development of radio techniques enabled the spectral line emission from neutral hydrogen to be studied it was possible to make only rough estimates of the amount of hydrogen in the galaxy. However during the last two decades these new techniques have given more precise information about the distribution of gas in the galaxy, and it is now believed that only about 2 percent of the mass of the galaxy is in the form of gas and dust. Most of this interstellar material is in the form of hydrogen gas, with a small amount of dust—about 1 percent of the mass of the gas. There is a wide variation in the distribution of this interstellar material. In the spiral arms in the general vicinity of the Sun the gas comprises about 20 percent of the total mass. In the regions closer to the center the distribution changes significantly. Within 2,000 light years of the galactic center the total mass is 10^9 $M_\odot$, but less than 1 percent of this is in the form of gas. Recent work has also shown that within 100 light years of the center there is a condensation of stars with a total mass of 2×10^8 $M_\odot$. Again the gas contributes less than 1 percent of the total mass, and in this central core at least one half percent of the mass appears to be in the form of ionized hydrogen gas.

The galaxy therefore shows a striking variation in the distribution of mass between the stars and the interstellar gas and dust as the distance from the center increases. Broadly speaking, in the regions within 2,000 light years of the center the mass is largely in the form of stars, whereas in the spiral arms about one-fifth of the mass is in the form of interstellar material. This difference is exemplified also in a marked difference between the types of stars in the central region and those in the spiral arms. In the spiral arms the stellar population is predominantly of young blue stars (Population I), with stars still condensing from the interstellar gas clouds. On the other hand, in the central regions the stars are predominantly old red stars (Population II), and star formation is no longer taking place.

The reason for these differences in gaseous content and age of stars is a subject of contemporary speculation. Any theory of the evolution of the galaxy must explain why nearly all the gas in the central regions of the galaxy long ago formed into stars. One idea, for example, is that the original gaseous cloud of primeval hydrogen from which the galaxy formed was not of uniform density but had a dense central region—the rate of star formation being proportional to the square of the density of the gas. Another view is that the spiral arms were ejected from the nucleus of the galaxy at a late stage in its formation.

THE LIFE CYCLE OF STARS

The Birth of Stars

Ideas of how galaxies like the Milky Way are formed will be discussed again in Chapter 4. These views are largely speculative, because there seems little chance that it will ever be possible to make observations of the formative process. The situation is quite different for the stars. In recent years we believe that observations have been made both of the birth and of the death of individual stars.

Once the galaxy has condensed from the primeval gas the possibilities exist for subsequent condensation into stars. Given regions of sufficiently high gas density in a primordial galaxy, the random motions of the atoms give rise to condensed pockets of gas containing so many atoms that each condensation is preserved by its own gravitation. The number of atoms necessary to give rise to a self-gravitating mass of gas that will eventually form a star similar to the Sun is very large—of the order of 10^{57}. (For comparison the number of neutrons and protons contained in all the atoms that make up the Earth is only about 10^{51}.)

This globule, or protostar, continues to contract under its own gravitation, and as it does, the release of gravitational energy increases the temperature of the protostar. Originally the temperature of the contracting mass is the same as that of the gas cloud from which it was formed, say about 100 K. Calculations indicate that in a very short time (about a year) the temperature at the center of such a contracting gaseous mass increases to about 50,000 K. The hydrogen atoms become ionized, so that the central region now consists of protons and electrons. Further, the surface of the protostar (which has decreased in diameter during this time from trillions of miles to a few hundred million miles) is at a temperature of the order of 1,000 K and is luminous. The process of condensation and increasing temperature continues rapidly. After a few more years, when the protostar has contracted to a diameter of about 30 million miles, its internal temperature has risen to about 150,000 K and its surface temperature to about 3,500 K. The protostar is then a star in the sense that it has become a highly luminous red object, much cooler than the Sun but with a diameter 50 times greater.

A condensing globule of this size reaches a new phase about 10 million years after the beginning of the condensation. Its diameter is then about 1.5 million miles, and the temperature in its interior has risen to several million degrees so that thermonuclear reactions have commenced. Initially the fusion reaction is that in which the collisions of two protons yield deuterons. The deuterons then undergo fusion reactions with protons to form the light isotope of helium, helium 3. With further contraction the temperature continues to rise until the main thermonuclear reactions involve the conversion of hydrogen to helium 4. When a star reaches this stage, about 27 million years after the beginning of the condensation, it reaches a stable condition in which the pressure generated by the release of nuclear energy prevents further contraction due to self-gravitation.

In the example chosen the condensing globule has become a star with about the same mass and luminosity as the Sun. The interior of the Sun is at a temperature of about 20 million degrees and at a pressure of several thousand million atmospheres. The thermonuclear transformation of hydrogen to helium is taking place at the rate of 564 million tons of hydrogen to 560 million tons of helium per second; the 4 million tons of matter converted to energy each second gives an output of 4×10^{23} kilowatts. Since the mass of the Sun is 10^{27} tons it has a life expectancy in this stable condition of several thousand million years.

The Sun is a star of average size and luminosity, with a surface temperature of 6,000 K. Sixty years ago Hertzsprung and Russell discovered the important relationship between the temperature (and spectral type) of stars and their luminosity (or absolute visual magnitude). They found that the majority of the stars lie on a clearly defined band from the hot blue giants to the cool red dwarfs as shown in Figure 3.1, the Hertzsprung-Russell, or H-R, diagram. This band is known as the main sequence—it is the region where stars have the stable configuration with outward pressure resulting

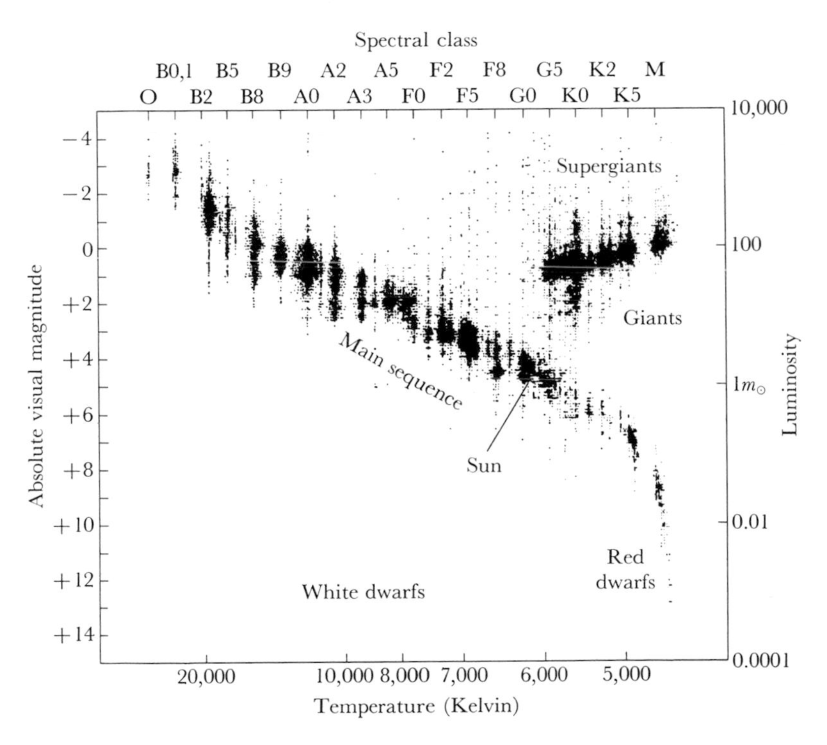

FIGURE 3.1. *The Hertzsprung-Russell diagram, based on the observations of 6,700 stars. The ordinates are absolute visual magnitude (left) and luminosity in terms of the Sun as unity (right). The abscissae are spectral class (top) and surface temperature of the star (bottom). (From the diagram originally drawn by W. Gyllenberg of Lund Observatory.)*

from thermonuclear processes in the deep interior balancing the self-gravitating forces. The initial size of the contracting protostar determines where the star arrives on the main sequence. The most luminous and bluest stars at the top left of the main sequence have masses about 60 times greater than the Sun. At the bottom right of the main sequence low-temperature red dwarfs have been observed with masses less than one-twentieth of the mass of the Sun. For a much smaller mass of the contracting globule, calculations indicate that the protostar never reaches the stage of hydrogen burning in thermonuclear processes; it never reaches the main sequence and eventually becomes an unobservable black dwarf. Contrary to what might be expected, the evolutionary processes for the large stars are much faster than for the dwarfs. A star ten times the mass of the Sun has a 100 times shorter life time, because the greater mass leads to higher temperatures and a disproportionate increase in reaction rates—a doubling of the temperature implies an increase in the reaction rates by 30,000 times. The converse is true for the dwarfs.

This description of the formation of stars from primeval gas clouds is based on calculations of the theoretical behavior of a contracting globule of gas; we may inquire if any observational evidence exists to support these concepts. An affirmative answer is possible. Figure 3.2 shows a gas cloud in the Milky Way in which large numbers of dark globules are discernible. These are believed to be protostars in the early stages of condensation, which, in a few tens of millions years, will reach the main sequence. As a striking example, Figure 3.3 compares two photographs of the same region of the Orion nebula taken in 1947 and 1954, showing the emergence in a time of seven years of a fourth protostar in this region. Further evidence has been obtained from infrared studies of the sky. In recent years many objects have been observed that radiate in the infrared, but not in the visible, region of the spectrum, precisely as would be anticipated for the earliest low-temperature phase of a condensing protostar.

From the study of the spectra of the stars it is possible to make estimates of the relative abundances of the elements in the universe.

FIGURE 3.2. *The gaseous nebula in Orion, typical of the gas clouds in the Milky Way in which protostars can be seen in process of formation. (Hale Observatories)*

As expected, hydrogen predominates, to the extent of 66 percent; there is about 30 percent of helium and only a few percent of all the other elements. Given hydrogen (or protons and electrons) as the primeval material of the universe, how did the heavier elements

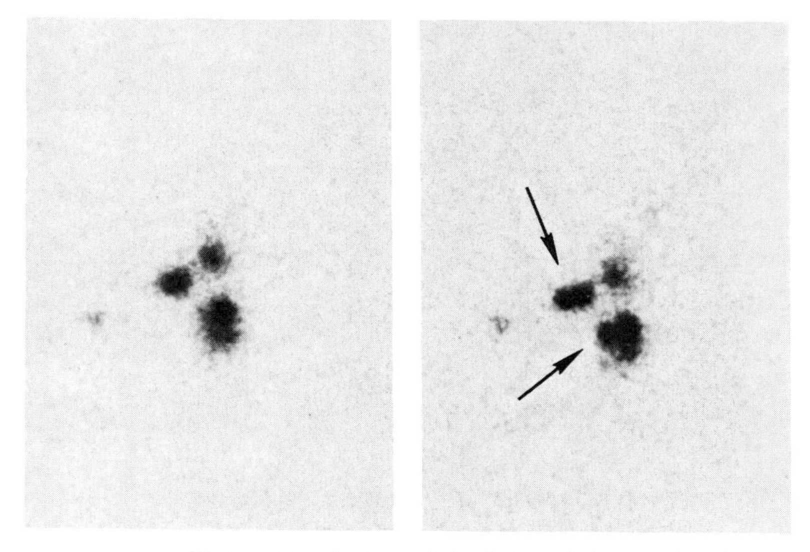

FIGURE 3.3. *Photographs of a part of the Orion nebula, showing the formation of protostars. The left-hand photograph was taken in 1947 and the right-hand photograph in 1954. In this short time a fourth protostar can be seen to have emerged from the gas cloud.* (*Lick Observatory*)

arise? In G. Gamow's theory (see Chapter 4) they were all formed in the first few minutes after the beginning of the expansion of the universe. Later it was shown that this theory of the build-up of the elements was untenable, and that in the initial minutes of the universe the build-up of the elements could have reached only as far as helium. Many astrophysicists now believe that the observed helium content of the universe was formed from the primeval material in the early stages of the universe. Our present beliefs about the formation of the elements heavier than helium are largely based on the work of F. Hoyle, W. Fowler, and G. and E. M. Burbidge. In this theory the heavy elements are formed by thermonuclear reactions in the interiors of the stars on the main sequence. Many problems of detail remain unsolved, especially the question of the abundance of helium and the time of its formation; but the main theoretical and observational framework, from the initial condensation to the arrival of stars on the main sequence, seems to be reasonably well established.

The Death of Stars

White Dwarfs. Although the condition of a star becomes one of long-term stability once it attains the hydrogen-to-helium thermonuclear stage on the main sequence, it is evident that it cannot have an infinitely long life. Eventually the supply of hydrogen will be exhausted (another few billion years in the case of the Sun). The investigation of this final phase of stellar evolution is one of the most extraordinary aspects of contemporary astronomy. Theoretical calculations of the behavior of the star as it reaches the end of the hydrogen-burning phase show that the star begins to contract because the gravitational forces are no longer balanced by energy production in the interior. The core of the star is now composed mainly of helium, which heats up as the star contracts, so that eventually the temperature becomes high enough in the outer gaseous regions of the star for the hydrogen there to undergo thermonuclear transformation to helium. The star moves over to the right of the main sequence because, although it is expanding, its luminosity does not change significantly, so that its surface temperature diminishes. For a star like the Sun this phase of hydrogen burning in the shell may last for 5 or 6 billion years.

Subsequent events depend critically on the mass of the star, as we shall see. In general the hydrogen burning of the shell eventually becomes so great that, although the star's luminosity increases, its diameter also increases to the extent that its surface temperature remains nearly constant; the star moves upward on the H-R diagram to become a red giant. The continued contraction of the core of helium eventually causes its temperature to rise to 100 million degrees, at which point commences a sequence of processes involving the thermonuclear transformation of helium to carbon in the core. These processes may cause the star to change its luminosity, size, and surface temperature so that it oscillates one or more times towards the main sequence. During this period the conditions in the core are computed to be such that the helium reaction may become an explosive one, known as the helium flash. Eventually the core becomes largely transformed to carbon, with helium now burning in the shell. Finally all further possibilities of sustaining the stellar envelope at a great size by nuclear burning come to an end and the star collapses, crossing the main sequence to the left.

This is the particularly fascinating point in the history of the collapse of a star. There is a fundamental principle in atomic physics known as the Pauli exclusion principle, which states that no two electrons in an atom can occupy the same energy state. In a star like the Sun, the temperature of the core of carbon never gets high enough for carbon burning to take place, and the star's collapse continues until all the electrons are packed together as closely as the exclusion principle allows; at that stage an equilibrium is reached. This occurs when the star's diameter is reduced to only a fiftieth or a hundredth of that of the Sun—say 15,000 to 30,000 kilometers (the diameter of the Earth is 12,600 kilometers). The density is very high—of the order of 6×10^5 grams per cubic centimeter (say 10 tons in a cubic inch). Because of its small size such a star, called a white dwarf, is very faint, although its surface temperature may be 10,000 K. Once this white-dwarf stage is reached the star shrinks only slowly. Eventually the remaining heat radiates away, and the star becomes unobservable as a black dwarf.

Supernovae, Neutron Stars, and Pulsars. This relatively peaceful end to the life of a star is now believed to take place only for stars that are not much more massive than the Sun. In 1931 Chandrasekhar calculated that there must be a limiting size above which, notwithstanding the exclusion principle, a white dwarf could not sustain itself against further collapse. He calculated that if a star was 1.4 times as massive as the Sun the gravitational force during the collapse would be so great that the nuclear forces, which normally determine the atomic arrangements according to the exclusion principle, would be overridden and the electrons and protons in the core would be forced together to form neutrons. The end result would be a neutron star, in which all the material of the original collapsing star would be packed into a sphere only about 10 kilometers in diameter—that is, only a three-thousandth of the diameter of a white dwarf, but with a density a hundred million times greater. In the neutron star the exclusion principle, which normally applies to the energy states of the electron, now applies to the neutrons, so that further collapse is prevented. Modern computations have reduced the Chandrasekhar limit somewhat, from about 1.4 to 1.2 times the mass of the Sun. With such a mass contained within a sphere of diameter 10 kilometers or so, the density is such that a spoonful of the material would weigh billions of tons.

A collapse to this condition does not occur gradually, like the final stages of the collapse of a star to a white dwarf. The greater mass of material increases the temperature of the contracting carbon core to 600 million degrees, at which temperature thermonuclear reactions involving the carbon core are initiated. A sequence of events ensues leading to the production of heavier elements; further contraction and higher temperature continue until iron is produced. The final collapse of the iron core creates temperatures of trillions of degrees and pressures of a trillion trillion tons per square inch, and the star explodes as a supernova. About half the mass of the star is blown out into space, and nuclear reactions involving the capture of neutrons and protons by iron lead to the formation of heavier elements. The time during which elements heavier than iron can be formed in this way is brief, and this is compatible with the sudden drop-off of 100,000 times in the abundance of elements heavier than iron. The remaining half of the stellar mass collapses finally into a neutron star.

Although these general concepts have been accepted for a considerable time, and the supernova explosion is a well-observed event, the existence of the remaining part of the star as a neutron star was nothing more than a theoretical postulate until a few years ago. In February 1968 A. Hewish and his colleagues at Cambridge published the surprising story of the discovery by his research student Jocelyn Bell of pulsating radio sources from space. The first source discovered emitted pulses about 300 milliseconds long every 1.337 seconds with extreme regularity; soon other similar pulsating sources were recognized. Investigations indicated that the emissions originated from objects at stellar distances, and since there were no obvious optical counterparts it was immediately suggested that the radio pulses had their origin in neutron stars. Less than a year after this announcement this suggestion received support when a pulsar in the Crab nebula was discovered, and further, a star near the central region of the nebula was found to be pulsating optically at the same rate as the radio pulses. By 1970 X-ray and γ-ray pulses coincident in phase with the radio and optical pulses had been found, thus extending the observations over a wavelength range of 10^{13} to 1. These discoveries seem to tie together in a remarkable manner the whole theory of the final collapsing phase of a star that is heavy enough to end its life as a supernova, with the production of a neutron star as the end product.

At this time (1974) just over 120 pulsars have been discovered by their radio emission. It may be somewhat surprising that only one or two other pulsars have been satisfactorily related to visible supernova remnants. However, it is likely that the catastrophic explosion of the supernova ejects the remnant neutron star at a high velocity from the position of the original contracting star, so that only young pulsars would be expected to be found in association with the gaseous remnant. This view is compatible with the fact that the Crab nebula pulsar and the other well-attested pulsar, associated with the Vela supernova remnant, have the shortest periods known—33 and 89 milliseconds respectively. Since the majority of the pulsars investigated show a very slow increase in period with time, it is believed that the emission is associated closely with the spinning of the neutron star, and that as energy is emitted, the rate of spin slowly decreases. Hence, those of longer period would be older and might well have moved far away from the visible remnant. It is hoped that attempts to measure the proper motion of the pulsars will substantiate this belief. Although the evidence appears to be satisfactorily in favor of the emission of the pulses from a spinning neutron star, it should be remarked that, at this moment, there is no agreed explanation of how the energy of the rotation of the star is converted into pulses of electromagnetic energy, which in the case of the Crab nebula pulsar extend over the entire spectral range.

Black Holes. According to modern computations a star only slightly more massive than the Sun will end its life in a supernova explosion and a neutron star. But stars are known that are at least fifty times heavier than the Sun. Will these follow the same pattern and end their life in the cataclysmic sequence of the supernova and the neutron star? This question, whether there is a limiting mass above which even a neutron star could not sustain itself against further collapse, was investigated in 1939 by J. R. Oppenheimer and G. M. Volkoff. Their answer was that, indeed, for certain masses the gravitational forces would become so enormous that the neutron star would suffer further collapse. The mass necessary to cause further contraction of a neutron star is uncertain, but it is widely assumed that a star with a mass more than 3 times that of the Sun would suffer this further stage of collapse.

Modern theory indicates that further collapse beyond the neutron star is not just a matter of increasing the density of the core, because there is no further stage of equilibrium. The collapse continues until the star passes inside the event horizon and becomes a black hole. A somewhat surprising result of the theory is that only a relatively small further contraction is necessary. A star a little more massive than the Sun suffers a decrease in linear dimensions of 70,000 to 1 when it collapses to a neutron star. A red giant suffers a linear contraction of 20 million to 1. But only a further linear contraction of about 3 to 1 appears to be necessary to take the remnant inside the event horizon, where the gravitational forces are so immense that no signals can escape.

Because no signals can escape from a black hole it may appear that the concept could never be more than a theoretical speculation, since the black hole would be unobservable. It is a surprising fact that there is now believed to be observational evidence for black holes. One of the many X-ray sources investigated by the equipment in the UHURU satellite was the source known as Cyg X-1, one of the four X-ray sources in the constellation of Cygnus discovered in earlier rocket flights. In the case of Cyg X-1, the UHURU observations in 1971 revealed that the X-ray emission was pulsating at the rate of several times per second. Shortly afterwards radio astronomers in Holland and America discovered that Cyg X-1 was a weak but variable radio emitter. They measured the position of the source so precisely that observers at the Royal Greenwich

Observatory were able to identify Cyg X-1 optically. They found it to be a binary star with a period of 5.6 days, the central object being a 9th-magnitude supergiant. An estimate of the mass of this star at 12 solar masses and of the binary component at 3 solar masses was compatible with the behavior of the binary. Subsequent calculations placed the mass of the supergiant at 20 solar masses and of the component at 5 solar masses. There was one further important point. The X-ray emission undergoes large intensity changes in times as short as 100 milliseconds, which implies that the radius of the emitter must be less than one-tenth that of the Sun. An object of 5 solar masses but with only one-tenth of the solar radius can only escape gravitational collapse to the black hole condition if it is a rapidly rotating white dwarf, and even in that case the star would eventually suffer collapse. It is now believed that the companion of this supergiant star is a black hole, and that matter ejected by the supergiant is being swallowed into the black hole. As this matter streams beyond the event horizon of the black hole there is a large release of gravitational energy, and it is suggested that near or above the black hole a hot gas cloud is formed, in which X rays are produced with high efficiency.

THE INTERSTELLAR MEDIUM

Reference has already been made to the varying amounts of interstellar gas and dust in different regions of the Milky Way. Until the advent of radio investigations almost the only regions of the interstellar medium that could be observed were the clouds of hydrogen gas in the neighborhood of hot stars. In those regions (H II regions) the radiation from the stars ionizes the hydrogen gas. The subsequent energy transitions of captured electrons then emit light, and the gas clouds are observable as emission nebulae. Although from dynamical considerations there were believed to be considerable amounts of neutral hydrogen gas in interstellar space there was no means of confirming this.

This situation changed dramatically in 1951, when radio astronomers in Holland, America, and Australia succeeded, almost simultaneously, in observing a spectral line in the radio-wave region from neutral hydrogen in the galaxy. The suggestion that it should be possible to observe a spectral line from neutral hydrogen was made by H. C. van de Hulst during a colloquium in occupied Holland in 1945. The emission, on a wavelength of 21.1 centimeters, arises in the following way. The proton in the hydrogen atom has a magnetic moment, which causes the atom in the ground state to have two closely separated energy levels, depending on whether the electron is spinning in the same direction as the nucleus or in the opposite direction. When the spin of the electron reverses, the energy change is such that the resultant radiation is emitted with the wavelength of 21.1 centimeters. Although for any particular atom the transition occurs on the average only once in 11 million years, van de Hulst calculated that because of the great number of atoms in the neutral hydrogen clouds of the galaxy this spectral line should be detectable.

The successful detection of this spectral line in 1951 and the subsequent development of the technique has been a matter of very great importance to our understanding of the galaxy. The investigation of this neutral hydrogen gas (H I regions) made possible a quantitative assessment of the amount of gas in various parts of the galaxy. Perhaps of even greater significance is that, since the radiation is emitted at a precise frequency, any relative motion between the hydrogen gas and the observer is manifested as a Doppler shift in the frequency. Thus by measuring the frequency at which the line is actually observed from different parts of the galaxy it is possible to construct an accurate dynamical model from the relative velocities derived from these frequency shifts. So much of the galaxy is obscured from view by interstellar dust that hitherto the spiral structure could only be inferred. Now, since the radio waves penetrate the dust clouds without absorption, the detailed spiral structure was revealed for the first time.

The development of the special equipment needed for a search in radio frequency and position was later aided by the application of digital techniques and computer processing. These high-speed frequency-search and integrative techniques, coupled with the availability of highly sensitive electronic equipment in the centimeter range, stimulated the search for other molecular constituents of the interstellar medium. Success was achieved in 1963 when an American team discovered the spectral line of wavelength 18 centimeters from the hydroxyl radical (OH). At about this wavelength a number of lines arising from energy-level transitions in the OH molecule are predicted theoretically; four of these have been investigated extensively. Some of the OH regions have been found to have extremely small angular extent, and it is believed that a maser type of process must be involved in the emission from these clouds. In some cases OH emission has been detected from the regions of space where other evidence, already mentioned, indicates that star formation is occurring.

Since 1969 spectral lines from many other molecules in the interstellar medium have been detected, many of them at short wavelengths in the low centimeter and millimeter region. In the latter case observations have to be made by accurately figured radio telescopes located on high and dry mountain sites in order to avoid atmospheric problems. The spectral line in the 6-centimeter waveband from formaldehyde (H_2CO) was the first of these new lines to be observed, in 1969, and by the end of 1972 25 molecules in interstellar space had been detected by their radio spectral lines.

Clearly an entirely new branch of astronomy has been opened by this work. Although the molecular constituents make only a small addition to the mass of the interstellar hydrogen gas they are of great importance to the attempt to understand the processes at work in space, especially in the regions of the spiral arms where star formation is still occurring. The unexpected discovery of the existence of complex molecules, including water, in the gas clouds from which stars are evolving lends a new and vital interest to the speculations of exobiology about the evolution of life elsewhere in the Milky Way.

CHAPTER FOUR

THE EXTRAGALACTIC SYSTEM AND THE COSMOLOGICAL PROBLEM

A FEW YEARS AFTER Shapley's discovery of the extent of the Milky Way, astronomers reached even more profound conclusions about the nature of the universe when Edwin Hubble, working with the new 100-inch Mt. Wilson telescope, obtained definitive evidence that the Milky Way system of stars did not comprise the totality of the universe.

The existence of nebulae, bearing the appearance of luminous diffuse regions of gas, had long been recognized. In a paper, "The construction of the heavens," that Sir William Herschel read to The Royal Society in 1785 he explained why he found it necessary to build larger telescopes. He said that as the power of the telescope is increased the observer "perceives that the objects which had been called nebulae are evidently nothing but clusters of stars." However, with his 48-inch telescope Herschel found that he could not resolve all the nebulae into stars; many retained a diffuse gaseous appearance. He suggested that some nebulae were indeed gaseous objects within the Milky Way system, but that others—those that could be resolved into stars—were external to the Milky Way. Herschel was correct in this opinion, but since no definite observational evidence could be obtained his ideas remained speculation for over a century. Lord Rosse, with the much larger 72-inch telescope that he had built at Birr in Ireland in the middle of the 19th century, obtained much more detail about these nebulae, and the drawings he made clearly show their spiral formation.

In a famous "Great Debate" held in Washington in 1920, Shapley defended the results he had just published about the size of the Milky Way and the position of the Sun among the stars. He was opposed by Heber Curtis of Lick Observatory. On those issues Shapley was right, but on the other contentious issue of that time—whether the spiral galaxies were within the Milky Way or separate star systems at great distances from it—Curtis was correct. From measurements of their size and magnitude Curtis produced further evidence in support of Herschel's speculation. But once more the key to an unambiguous answer lay in the ability to measure distances.

In 1926 Hubble published the results of his observations on the nebula catalogued as M33. He had been able to resolve the nebula into stars, among which he identified Cepheid variables, and thus, as Shapley had done for the globular clusters, he was able to measure the distance to M33 by using the period-luminosity relationship. He thereby produced the first unambiguous proof that a nebula was an extragalactic star system remote from the Milky Way. In the same year he published his results on 400 of these extragalactic star systems. In his paper on M33 he referred to similar measurements on M31, the great spiral nebula in Andromeda (a naked-eye object under favorable conditions)—a nebula 2 million light years distant, which we now believe to be almost identical to the Milky Way in size, structure, and stellar content.

In the second of these papers on the statistics of 400 nebulae, Hubble made the assumption that nebulae of the same type would have the same absolute magnitude. From the distance measurements he had made on the nearer of these samples, in which it had been possible to identify Cepheid variables, he estimated that the 100-inch telescope when photographing to a limiting magnitude of +18 would encompass 2 million extragalactic nebulae, to a distance of 140 million light years. Further, he estimated that this distance represented about $\frac{1}{600}$ of the radius of curvature of the universe of general relativity, in which the total mass would correspond to about 3.5×10^{15} nebulae. He concluded that "with reasonable increases in the speed of plates and size of telescopes it may become possible to observe an appreciable fraction of the Einstein universe." Two typical extragalactic nebulae are shown in Figures 4.1 and 4.2. Our own Milky Way would have this appearance when viewed face-on (Figure 4.1) or edge-on (Figure 4.2) by a distant observer.

Clearly, in a few years a vast change had occurred in the concept of the nature and extent of the universe. A similar change had occurred, too, in the theoretical basis of cosmology, as evidenced by Hubble's reference to the Einstein universe. Hitherto it had been believed that the universe was effectively a static assembly of stars, and that any motions of the objects in the universe were small. It seemed to be impossible to explain the stability of such a universe on the basis of Newtonian theory, since there would be an attraction between stars obeying the inverse-square law. At the turn of the century a suggestion was made that the simple Newtonian law might not apply throughout all space and that the gravitational law should contain a term such that at large distances repulsive forces balance Newtonian attraction. Since there seemed to be no physical basis for such a concept, it is understandable that there was little interest in cosmology.

This situation changed suddenly in 1916 with the publication by Einstein of the general theory of relativity. In rejecting the Newtonian concept of absolute space, Einstein attempted in 1917 to express the principle that the behavior of one body could be considered only in relation to all other bodies in the universe. However, when Einstein applied the field equations of general relativity to this problem he was unable to choose boundary conditions such that the inertial field was fully determined by the masses in the

FIGURE 4.1. *The galaxy NGC 5457 in Ursa Major, seen face-on, photographed by the 200-inch Palomar telescope. (Hale Observatories)*

universe. He overcame this difficulty by introducing an additional term, the cosmological constant, that when positive gave a solution with uniform density of matter, random velocity zero, and space

FIGURE 4.2. *The galaxy NGC 4565 in Coma Berenices, seen edge-on, photographed by the 200-inch Palomar telescope.* (*Hale Observatories*)

curved so that it was finite but unbounded. The difficulties of the boundary conditions at infinity were abolished. There was no solution for empty space with the cosmological constant positive, and Einstein believed that he had found a unique solution predicting a static universe, and moreover, a solution that gave full expression to the identity of gravitational and inertial masses in the universe.

The few years following the publication of this theory were both curious and remarkable ones in the development of cosmology. On the theoretical side, within a few months of Einstein's publication Willem de Sitter showed that the Einstein solution was not unique and that in fact there was a solution describing a universe that was only static if empty of matter. Particles introduced into this universe would recede from one another with ever-increasing velocity. At this time, the nature of the Milky Way and then of the large-scale structure of the universe were being revealed through the work of

Shapley and Hubble, but it was still believed that the stars and galaxies were essentially a static assemblage in which any motions of the objects were at velocities small compared with the velocity of light.

The last of the great discoveries of this epoch soon followed, that is, the discovery of the cosmological expansion of the universe. The phenomenon of the shift in wavelength of the light from the nebulae toward the red end of the spectrum had been known for many years. By 1924 V. M. Slipher had made such measurements of 38 nebulae, but the cosmological significance of this redshift was not realized until Hubble obtained evidence for the extragalactic nature of these objects. Between 1929 and 1931 he published results revealing the relation between the distance of a nebula and its redshift. With the redshift interpreted as a Doppler shift, the results showed that the speed of recession of the nebulae increased linearly with their distance. At that time the measurements extended to a distance of 100 million light years, where the recessional velocity was 2,000 miles per second.

THE PRESENT STATE OF COSMOLOGY

Forty years ago astronomers achieved the first understanding of the large-scale structure of the universe. With its great light-gathering power the 200-inch Palomar telescope subsequently extended the penetration into space to about 2,000 million light years. As will be seen, developments in radio astronomy have now at least doubled and probably tripled this penetration. How many extragalactic nebulae lie within the field of view of modern telescopes? Some idea of the immense number can be deduced from the sky photographs now being produced by the United Kingdom's new 48-inch Schmidt telescope in Australia. The faintest object photographed by this instrument has a magnitude of +23 and about 2,500 extragalactic objects can be counted in a square degree of the sky.* There are

*R. J. Dodd, D. H. Morgan, K. Nandy, V. C. Reddish, and H. Seddon, *Monthly Notices of the Royal Astronomical Society 171,* 329, 1975.

41,250 square degrees in the whole sky; thus, to the 23rd magnitude there are about 100 million galaxies. At this sensitivity limit the number of observable objects is still increasing by 2 or 3 times per magnitude. With larger telescopes and modern instrumentation the sensitivity limit can probably be improved by about 2 magnitudes, giving some 500 to 1000 million extragalactic objects within the observational limits of modern telescopes on Earth. Reference has been made to the future telescopes that will make it possible to observe even fainter objects. (The estimated limit for the 120-inch large space telescope is 5 magnitudes fainter.) Will the number of extragalactic objects continue to increase at this rate of 2 or 3 times per magnitude? At least to that magnitude, most astronomers would probably answer in the affirmative with some confidence, but how long this rate of increase will continue is one of the fundamental unanswered problems in cosmology.

There is another deep and unsolved problem concerning the redshift measurements. Hubble established that there was a linear relationship between the redshift (velocity of recession) of a galaxy and its distance. The linearity of this Hubble law was subsequently extended to greater distances, and today there is general agreement that the linear relationship extends at least to redshifts indicative of recessional velocities of nearly half the velocity of light, where the distance involved is of the order of 4,500 million light years. This recession of the galaxies implies that the universe is expanding at a rate that increases as we penetrate farther into space. For how long does the rate of expansion increase linearly with distance?

THEORETICAL MODELS OF THE UNIVERSE

Because of the undetermined cosmological constant in the general theory of relativity, a wide range of theoretical models for the universe can be derived. They make, in common, a prediction that the radius of curvature and the mean density of the universe vary with time. Within this generality it appears that wide variations are

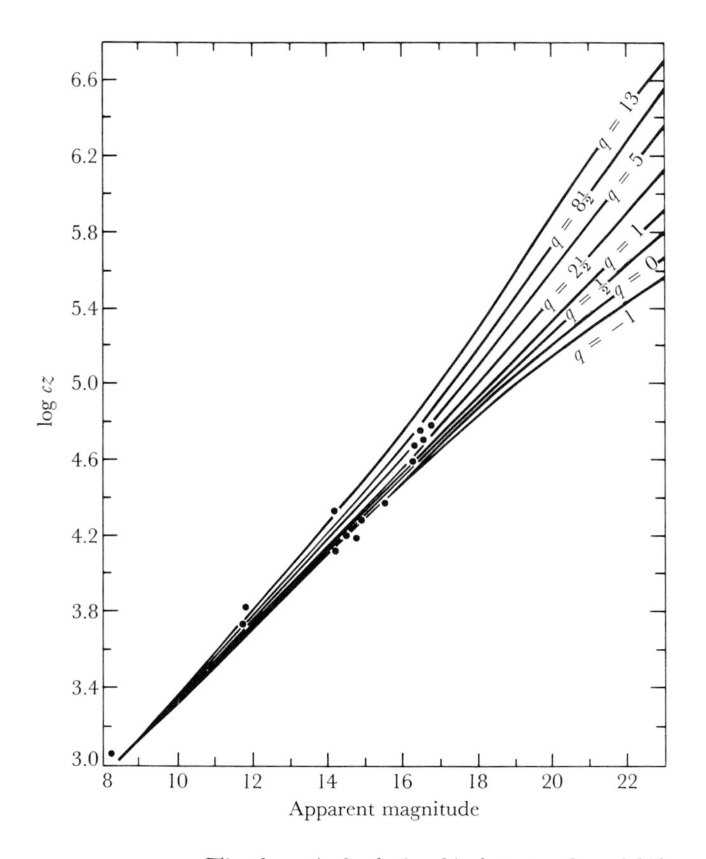

FIGURE 4.3. *The theoretical relationship between the redshift ($z = \Delta\lambda/\lambda$; the ordinate is logarithm of zc, where $c = 3 \times 10^5$ kilometers per second) and the apparent magnitude, for cosmological models with zero cosmological constant and various values of the deceleration parameter q. The points represent distant clusters of galaxies for which the relation was well established in 1961. The data as of 1974 are shown in Figure 4.11. (This diagram is based on the paper by A. Sandage,* Astrophysical Journal 133, *355, 1961.*)

theoretically possible—for example with some parameters the rate of expansion tends to zero as the radius of curvature tends to infinity, while with other parameters the universe is predicted to reach a maximum size, and then contract as time increases. Unfortunately for the possibility of reaching an easy observational solution, they all predict that there should be a linear relationship between distance and velocity, at least to recessional velocities approaching half the velocity of light. At greater distances and greater recessional velocities the various theoretical models differ in their predictions about the nature of the departure from the linear Hubble law. Figure 4.3 illustrates this for the simplest category, in which the cosmological constant term is taken to be zero. This diagram is a plot of the theoretical relationship between the redshift (equivalent to the velocity of recession) and the apparent magnitude of the galaxy (equivalent to the distance), for various values of the deceleration parameter q. This deceleration parameter is defined mathematically to express the relation of the rate of expansion of the universe with time.

The wide variation in the predicted behavior of the universe for this category of model universes may be illustrated by three typical cases with various values of the deceleration parameter. The theoretical case with which observations are most frequently compared is $q = +\frac{1}{2}$, known as the Einstein-de Sitter universe. In this case the variation with time of the scale factor of the universe (equivalent to the radius of curvature) $R(t)$ is of the type shown in Figure 4.4, where the rate of expansion tends to zero as R tends to infinity. If q lies between zero and $+\frac{1}{2}$ then the rate of expansion of the universe decreases monotonically to some positive value as R increases to infinity, as shown in Figure 4.5. However, if q is greater

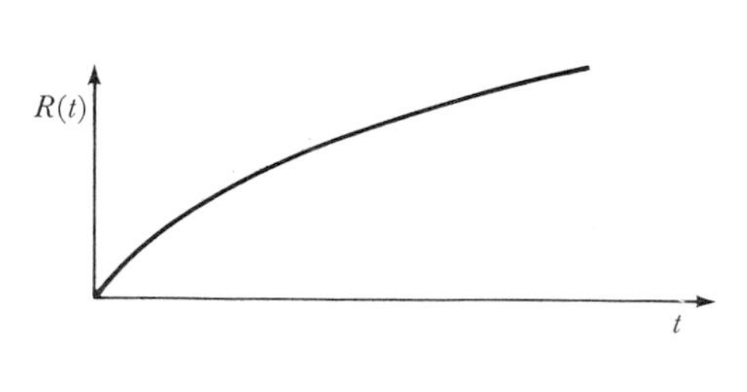

FIGURE 4.4. *The variation of the scale factor for the universe as a function of time, for the theoretical model with the cosmological constant zero and deceleration parameter $q = +\frac{1}{2}$ (the Einstein–de Sitter universe). (From B. Lovell,* Out of The Zenith, *Oxford University Press, 1973.)*

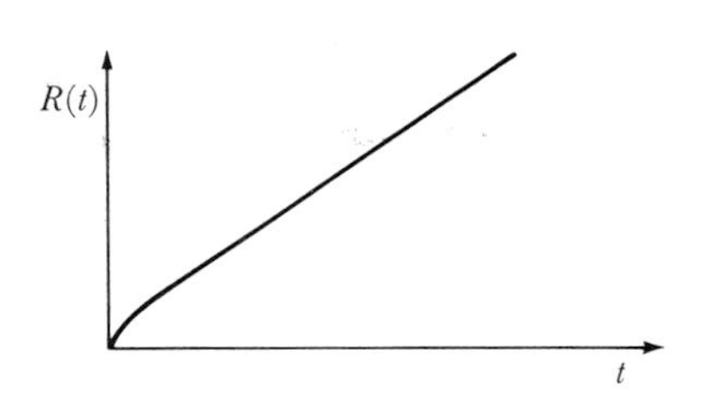

FIGURE 4.5. *The variation of the scale factor for the universe as a function of time, for the theoretical model with the cosmological constant zero and deceleration parameter lying between 0 and* $+\frac{1}{2}$ *(From B. Lovell,* Out of the Zenith, *Oxford University Press, 1973.)*

than $+\frac{1}{2}$ the universe is predicted to reach a maximum size and then contract as time increases, as shown in Figure 4.6.

It will be seen that, although the behavior of the universe as time increases is predicted differently, the models all have in common the implication that at time zero the universe must have been highly condensed. The same generality applies even to other categories of models wherein the simplifying assumption of a zero cosmological constant is no longer made. That is R and t have a common origin at the zero point. The mathematical and conceptual difficulties involved in this beginning of the universe from a singular condition at time zero stimulated the emergence in 1948 of the theory of the steady state. On this theory, the universe is unchanging when considered on the large scale, and it possesses a high degree of uniformity in time and space. Since the universe is expanding, this principle implies that new matter must continually be created in order to maintain a constant density. As the nebulae move apart,

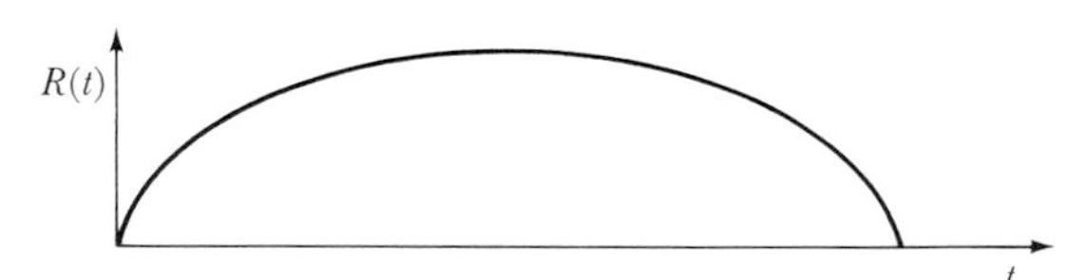

FIGURE 4.6 *The variation of the scale factor for the universe as a function of time, for the theoretical model with the cosmological constant zero and deceleration parameter greater than* $+\frac{1}{2}$. *(From B. Lovell,* Out of the Zenith, *Oxford University Press, 1973.)*

new ones are formed from the created matter at a rate sufficient to present an unchanging aspect to an observer at any point in time and space. The value of the deceleration parameter q in Figure 4.3 appropriate to the steady-state theory is $q = -1$.

OBSERVATIONAL TESTS OF THE THEORETICAL MODELS

For the last quarter of a century a major astronomical interest has been the attempt to make some observations on the universe that would enable a decision to be reached among these various conflicting theoretical model universes. The main avenues of investigation are as follows:

Number Counts

The question was raised earlier how far the number of objects would continue to increase with fainter magnitudes. As far as observations with optical telescopes are concerned, no change in the large-scale uniformity of the universe has been found to the greatest distance of penetration. If the universe has an infinite past history, as in the steady state, then no change is to be expected. On the evolutionary models, the number of objects per unit volume of space must have been greater in the early history of the universe. Further, since penetration into space implies the ability to look back into the past state of the universe, any evolutionary change in the universe is, in principle, observable. The difficulty is that it is necessary to achieve deep penetrations before the difference in the predictions become significant. The time elapsed since the beginning of the expansion in the evolutionary models can be calculated if the slope of the Hubble line connecting velocity of recession with distance is known.

The value of this Hubble constant has been subject to many revisions. The most recent authoritative figure, published in 1972 by Sandage,* is 55 kilometers per second per megaparsec. For a deceleration parameter q equal to zero, this corresponds to a time of 17,000 million years since the beginning of the expansion. For a value of $q = +1$, the corresponding time is 10,000 million years. On any of the plausible evolutionary models, the light from a galaxy with a redshift indicating a recessional velocity approaching half the velocity of light has taken about half of the time since the beginning of the expansion to reach us. Only at that distance are the differences in the predictions of various theoretical cosmologies expected to become significant.

Until the development of radio astronomy, it was not believed that such penetrations into the past history of the universe could be achieved. However, in 1951 the possibilities for exploration to cosmologically significant distances were completely changed by the discovery of hitherto unknown types of objects in the universe. The initial surveys of the radio emission from the sky revealed a general distribution of radio intensity, which was thought to arise in the interstellar hydrogen gas. As the sensitivity and resolution of radio telescopes improved, a number of localized sources of emission superimposed on this background were found. For some years these were believed to be objects within the Milky Way, especially since the third strongest of the sources coincided with a supernova remnant, the Crab nebula. No similar objects were obviously related to the other strong radio sources. In 1951 W. Baade and R. Minkowski succeeded in photographing one of these radio objects, in the constellation Cygnus. Their long exposure with the 200-inch telescope revealed a faint object of a type not hitherto recognized (Figure 4.7). The redshift of its spectral lines indicated that it was not in the Milky Way, but, on the contrary, was an extragalactic object at a distance of 700 million light years. In the following years many more radio sources were identified with distant extragalactic objects, and by 1959 an object had been identified, 3C 295 in Boötes, with a redshift implying a recessional velocity of more than

*A. Sandage, *Quarterly Journal of the Royal Astronomical Society 13,* 282, 1972.

FIGURE 4.7. *The peculiar radio galaxy in Cygnus, photographed by the 200-inch telescope. (Hale Observatories)*

40 percent of the velocity of light and a distance of 4,500 million light years.

Initially these objects were believed to be galaxies in collision, or interacting in some manner that produces relatively strong radio emission, although they are optically faint. They became known as radio galaxies, and are no longer believed to be colliding galaxies, but evidence of a violent disruption in the nucleus of a galaxy. In any case, it became evident that the localized sources of radio emission were distant objects in the universe, and on a statistical basis—without individual identification—it was argued that the distribution of their numbers with decreasing strength of radio emission would be equivalent to a count of the objects at increasing distances in the universe.

These counts of radio sources, made by groups of radio astronomers in the northern and southern hemispheres, led to intense arguments about their meaning. The most recent results published by the Cambridge group in England are shown in Figure 4.8, which is a plot of several thousand radio sources. The ordinate is the logarithm of the number (N) of radio sources per unit solid angle whose flux density (S watts per square meter per hertz) exceeds the value whose logarithm is plotted on the abscissa. This curve is compared with the straight line with a slope of $-\frac{3}{2}$, which would be the relation between log N and log S for a static universe containing a uniform distribution of sources. Now all the plausible theoretical models so far mentioned for an expanding universe predict a log N/log S curve with a slope less steep than $-\frac{3}{2}$. It appears that the only reasonable argument that would bring the observed slope more into alignment with the predicted slope is to assume that evolutionary effects have occurred in the sources themselves—in particular, that the objects with low values of S are the most distant, and are observed at an earlier phase in their evolution, when they had greater luminosity. Any arguments of this kind are, of course, incompatible with the steady-state theory, which makes a unique prediction of a log N/log S curve with a slope less steep than $-\frac{3}{2}$. Nevertheless, the implications and, indeed, the validity of the curves have been strongly contested by the adherents of the steady-state theory, mainly on the grounds that the counts are not of a homogeneous sample of objects and that the slope of the curve is strongly influenced by selection effects in the observations, especially the relatively small sample of sources available at the largest values of log S.

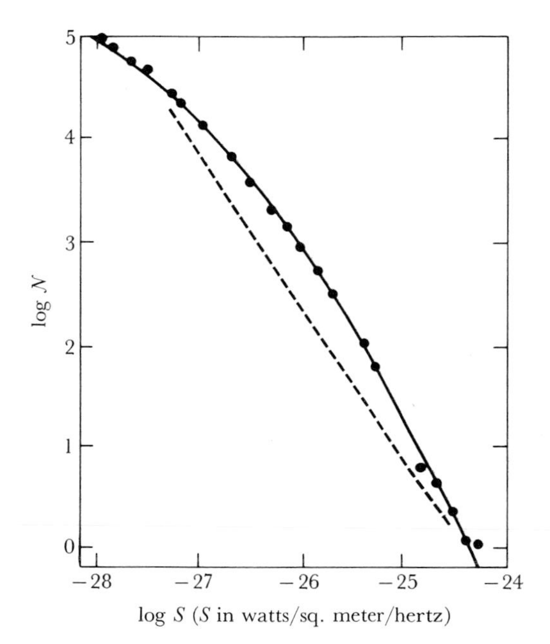

FIGURE 4.8. *The number counts of radio sources as a function of their intensity. The broken line has a slope of $-\frac{3}{2}$. The ordinate is the logarithm of the number N of sources per unit solid angle whose flux density exceeds the value S, plotted as log S on the abscissa. (Data from the paper by C. G. Pooley and M. Ryle,* Monthly Notices of the Royal Astronomical Society 139, *515, 1968.*)

The Microwave Background Radiation

While the debate on the interpretation of the counts of radio sources was in progress an entirely unexpected discovery was made that was immediately hailed as a decisive indication that the universe had evolved from a superdense condition. During 1964 and 1965 A. A. Penzias and R. W. Wilson of the Bell Telephone Laboratories, New Jersey, were testing new receiving equipment on a wavelength of 7 centimeters, intended for communication tests with the Echo balloon satellite. They found that when their aerial system was directed at the sky the noise level in the receiving equipment was 100 times greater than they had expected from the known sources of radio emission. Further, they found that the intensity remained constant from all directions, and in 1965 they announced that they had discovered an isotropic background radiation equivalent to a black-body temperature of 3.5 K. The measurements were soon confirmed by other scientists and extended to neighboring wavelengths. The effect became known as the 3-degree background radiation, although the most recent measurements correspond to a black-body curve for 2.7 K.

Strangely coincident with this discovery, Professor R. H. Dicke of Princeton, New Jersey, had been considering the implication of a high-temperature collapsed phase of the universe. He concluded that the temperature at such an epoch would have been 10^{10} K and that the relic black-body radiation should now be detectable at a temperature of 40 K. In fact, G. Gamow had proposed in 1946 that the universe began with a "hot big bang" and that relic radiation should now be observable at a temperature of 25 K. Penzias and Wilson assumed in 1965 that they had detected this relic radiation. Although the effective temperature is nearly 10 times lower than Gamow's prediction, the observation and theory are not necessarily incompatible, because the calculations are sensitive to the assumptions made about early helium production.

This discovery and its interpretation in terms of the relic radiation naturally led to a great debate, particularly since any conclusions in these terms would compel the abandonment of the theory of the steady state. An origin of this radiation in galaxies having a high output of energy at short radio wavelengths was suggested, but because of the high degree of isotropy of the radiation this has not appeared to be the solution. Other critics doubted if the measurements were those to be expected from black-body radiation. Although down to wavelengths of about 3 millimeters the measurements fitted the theoretical curve precisely (Figure 4.9; compare also the theoretical curves in Figure 1.1), the crucial question was whether the intensity fell off at even shorter wavelengths, since for this temperature the peak in the intensity of the radiation should occur at a wavelength of 1.5 millimeters. Unfortunately, there are

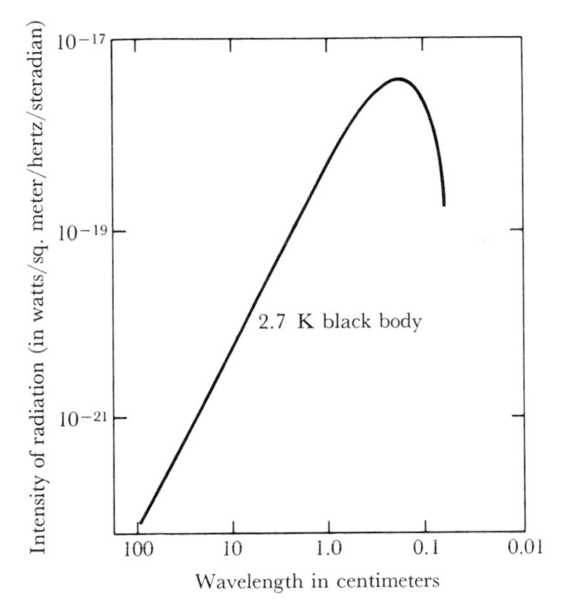

FIGURE 4.9. *The theoretical curve for the intensity of radiation as a function of wavelength for a black body at 2.7 K (compare with Figure 1.1).*

great difficulties in making this measurement, because of absorption of the radiation in the atmosphere, and also because several interstellar molecules have line spectra in this wavelength range. Several disparate results were obtained at wavelengths in the neighborhood of the predicted peak, but early in 1973 measurements were made down to a wavelength of 0.4 millimeter by equipment carried in a rocket launched from the Hawaii Test Range. The results confirmed the fall-off in intensity at wavelengths less than 1.5 millimeters, and were consistent with the concept that the measurements did indeed refer to a 2.7 K black-body radiation.

Details of a further experiment giving decisive confirmation were published in 1974 by British observers.* The observations were made with a Michelson interferometer, cooled by a liquid helium

*E. I. Robson, D. G. Vickers, J. S. Huizinga, J. E. Beckman, and P. E. Clegg, *Nature 251,* 591, 1974.

bath to a temperature of 1.4 K. It measured the spectrum of the background radiation from about 3 millimeters to a wavelength shorter than the predicted peak of the black-body curve. This equipment was carried in a balloon launched from Palestine, Texas, on 13 March 1974, and measurements were made for one hour at an altitude of 40 kilometers. These are the first direct spectral measurements of the background radiation in this critical region of the spectrum, and they show decisively a sharp decrease of the intensity at wavelengths shorter than 1.5 millimeters. The authors' conclusion is that, although there is a small discrepancy from the theoretical curve appropriate to a 2.7 K black-body radiation, it would be difficult to reconcile the measurements with a background temperature higher than 2.9 K. Therefore, it seems to be established that the radio emission discovered by Penzias and Wilson is black-body radiation at a temperature of about 2.7 K. At this moment (1974) no other realistic suggestion for the origin of this radiation has emerged and it must be regarded as decisive evidence in favor of the evolutionary universe and, furthermore, as indicative of a high-temperature state of the order of 10^{10} K in the initial moments of the expansion.

The Hubble Line at Large Redshifts; Quasars

The weight of evidence both from counts of radio sources and the microwave background radiation is that the universe has evolved from a dense initial state. If this interpretation is accepted, the major problem is then to determine the nature of the evolutionary universe—for example, which of the curves in Figures 4.4, 4.5, and 4.6 is appropriate. Will the universe continue to expand, or will the expansion eventually cease, followed by a contraction as in Figure 4.6? A straightforward answer could be obtained if it were possible to determine the mean density of the universe. For example, the Einstein-de Sitter universe (Figure 4.4) gives a definite prediction that the present mean density of the universe should be 2×10^{-29} grams per cubic centimeter. For Figure 4.6, the model that predicts an eventual contraction, the density must be greater than this value so that the gravitational forces will eventually predominate and lead to the collapsing phase. For Figure 4.5, in which the expansion remains finite as the radius of curvature approaches infinity, the density must be less than the figure given above.

Unfortunately, there is no known means of measuring the present density of the universe. The smoothed-out value of all the matter in the galaxies is believed to give a density of less than 10^{-30} grams per cubic centimeter, but there is no realistic method yet known enabling a sufficiently accurate measurement to be made of the matter that probably exists in intergalactic space.

The main contemporary approach to the problem, therefore, is the attempt to determine the value of the deceleration parameter q. If a precise value of q could be measured, a clear choice could be made from among the various possible categories of evolutionary models. The relation between radio-source counts and distance could, in principle, be used for this purpose. However, the interpretation of the existing radio log N/log S relationship already discussed appears to be influenced by a number of contentious issues, and there seems little hope that these measurements can be refined to determine the value of q in the foreseeable future. Attention is therefore concentrated on the relationship between the redshift (interpreted as a cosmological expansion of the universe) and the magnitude of objects (interpreted as distance). The theoretical curves have been given in Figure 4.3 for various values of q.

The difficulty of the problem may be illustrated by the case of a galaxy with a redshift of $0.5c$ (c = the velocity of light). The radio galaxy known as 3C 295 in Boötes discovered in 1959 has already been mentioned. It has a redshift of $0.461c$; hence at the time of its discovery astronomers had the ability to observe objects whose recessional velocities were of the order of half the velocity of light. Figure 4.3 shows that at this distance there is a predicted difference of 0.7 magnitude for a galaxy, depending on whether the universe is steady-state ($q = -1$) or Einstein-de Sitter ($q = +\frac{1}{2}$). Such small differences should be detectable by modern optical telescopes. However, the scatter in the results overrides these small differences, as illustrated by the points in Figure 4.3, which shows the results for distant clusters of galaxies whose redshift and magnitude values were well established in 1961.

Because of the much greater difference in the predictions for larger values of the redshift, it was natural that in the years following the discovery of the radio galaxy 3C 295 efforts were made to find even more distant objects. There was a most curious and exciting result of this search. A number of radio sources of even smaller angular diameter than the 3C 295 source were thought to be likely candidates. The immediate result of the search for optical counterparts with the 200-inch telescope was announced by Allan Sandage at the meeting of the American Astronomical Society in New York at the end of December 1960. In the position of one of these sources (3C 48) he had found a 16th magnitude star, unusually blue in color, with a faint wisp. He had found no identifiable spectral lines and concluded that the object was a "relatively nearby star with most peculiar properties." Two similar identifications were made shortly afterwards. Their stellar nature seemed to be confirmed by the observation of variations in the light output of the objects—it did not seem possible that objects as large as galaxies could show such variations over the short time periods observed.

Until 1963 the stellar nature of these blue objects was largely accepted. Then in March of that year Maarten Schmidt of Palomar published an identification of the spectral lines observed in a similar object known as 3C 273 whose position and peculiar radio properties had been determined by C. Hazard and others, using the radio telescope at Parkes in Australia. He concluded that 3C 273 was not a star in the Milky Way, but a distant galaxy of hitherto unrecognized type, with a redshift of 0.158*c* (Figure 4.10). At the same time J. Greenstein and T. A. Matthews of Palomar published similar results for the object 3C 48, which showed a redshift of 0.367*c*. By the end of that year nine further radio sources had been identified with starlike blue objects, and two of them, with redshifts of 0.425*c* and 0.545*c* became the most distant objects known, lying at distances of the order of 5,000 million light years. A new class of distant extragalactic object had been discovered, and they became generally known as quasars.

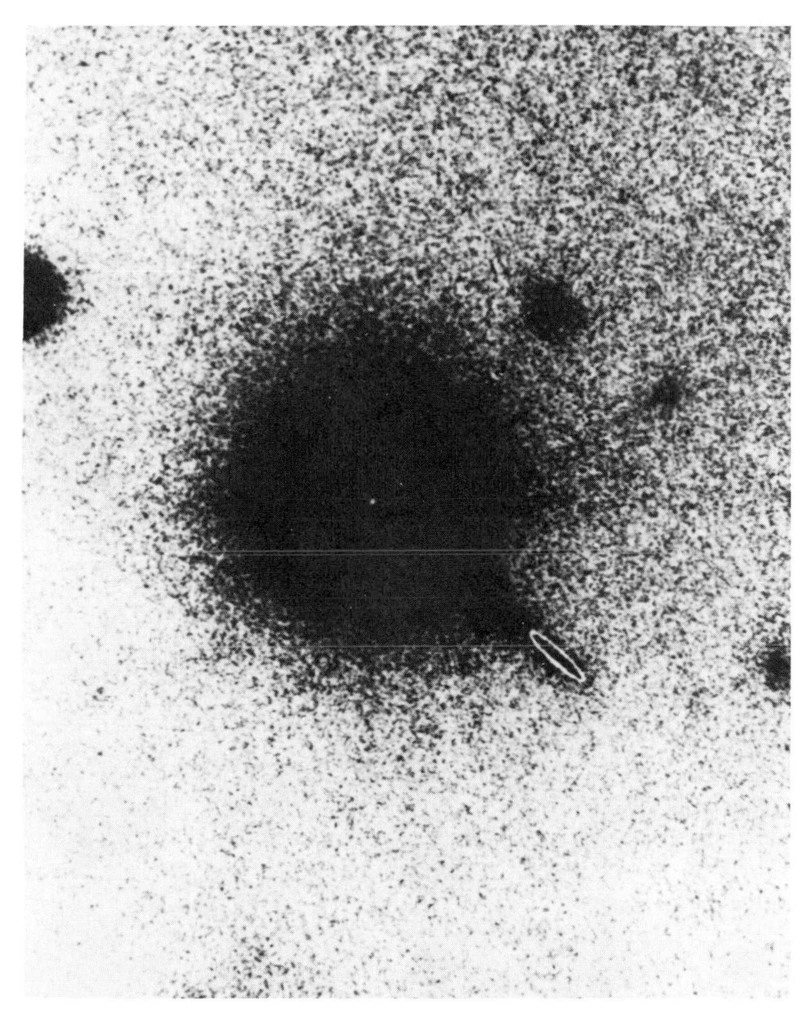

FIGURE 4.10. *The quasar 3C 273 in Virgo, overexposed to show the faint jet, outlined in white. The jet is 20 seconds of arc from the quasar, which is at the center of the image in the position of the white dot. (Hale Observatories)*

With the discovery of the quasars—relatively bright objects with high redshifts—it seemed likely that the Hubble line could be quickly extended so that the curvature of the universe could be obtained, with which to define a precise value of the deceleration parameter q. The hope that more quasars would be found with even greater redshifts certainly materialized. By 1970 redshift values z $(= \Delta\lambda/\lambda)$ of 2.8 had been found, indicating relativistic velocities of recession of more than 80 percent of the velocity of light, and in 1973 two quasars with redshifts of 3.4 and 3.5 were identified. At this time (1974) the redshifts of about 300 quasars have been determined (100 of these are radio quiet—high-redshift blue stellar objects without detectable radio emission).

The expectation that these measurements would define the deceleration parameter has not been borne out, as can be seen from Figure 4.11. The line on this diagram corresponds to a q value of $+1$. Whereas the clusters of galaxies and radio galaxies make a reasonable fit to this line—say to values log cz of about 4.8 (that is to redshifts of about $0.2c$)—the quasars of high redshift have introduced a great scatter. Indeed, to the present time the quasars have not contributed to any further understanding of the value of the deceleration parameter, although they have introduced many further problems into cosmological discussion.

The most significant of these are the doubts that have been raised about the interpretation of the redshift as a wholly cosmological phenomenon. There are many cases in which spectral lines are seen both in absorption and in emission from the same quasar, but they do not give the same value for the redshift. It has been suggested in these cases that the absorption is occurring in an expanding shell of gas surrounding the quasar, and that this interpretation could be consistent with a major redshift associated with the expansion of the universe. A greater difficulty is presented by a number of cases in which the quasar appears to be physically associated with a galaxy or group of galaxies of much lower redshift. There have been found recently a few examples of quasars in close proximity on the photographic plates but with markedly different redshifts. Of course, there is a definite chance that such associations will appear by coincidence on the two-dimensional image of the photographic plate, even though the quasars are at markedly different distances. The number of cases so far observed is too small to judge finally whether

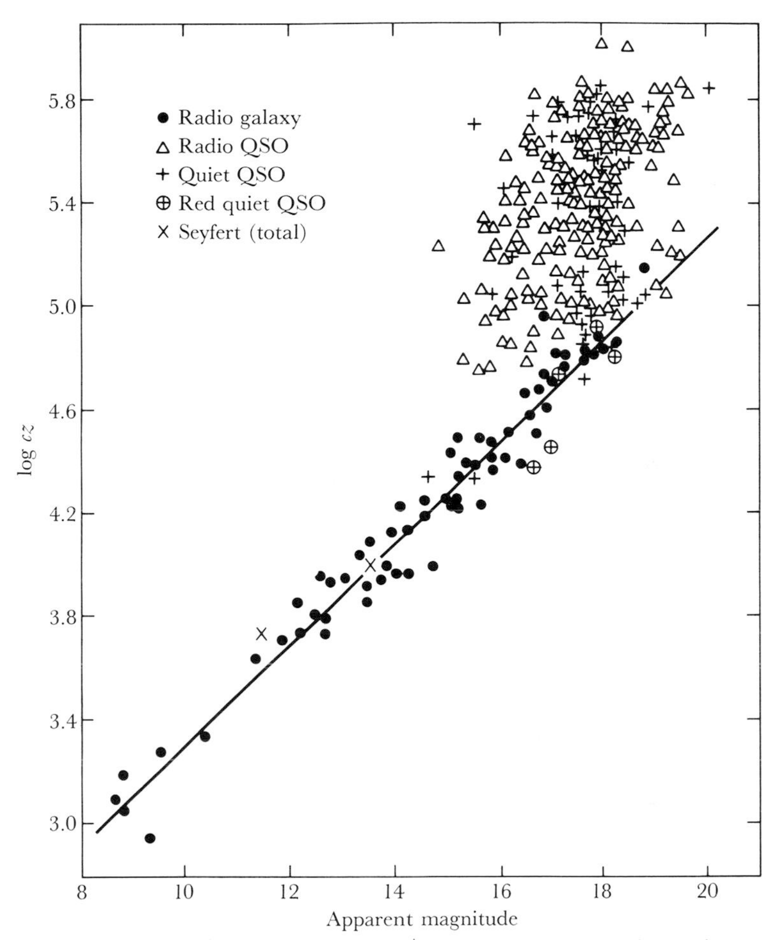

FIGURE 4.11. *The observed redshift/apparent-magnitude relation for radio galaxies, quasars, and other quasi-stellar objects as established to 1974. The line corresponds to a model with the deceleration parameter* $q = +1$. *The ordinate is the logarithm of cz with* $c = 3 \times 10^5$ *kilometers per second. Radio quiet QSO's are quasi-stellar objects without radio emission. Red quiet QSO's are quasars redder than the average quasar for a given redshift. (Adapted from the diagram given in A. Sandage,* Quarterly Journal of the Royal Astronomical Society 13, *282, 1972.)*

such associations are real or merely appear to be so because of chance associations on the plates.

If future investigations show that there are physically associated objects of markedly different redshift, then of course the cosmological interpretation of the redshift as an indicator of the expansion of the universe would no longer be valid. Other known causes leading to a redshift are actual high velocity, such as might result from a violent explosion, or a gravitational shift in wavelength arising because the spectral lines are emitted in a high gravitational field. Since it is difficult to conceive of these effects being compatible with other known features of the objects and at the same time making a significant contribution to the very high values of the redshift that are observed, some astrophysicists have been tempted to speculate that other, as yet unknown, effects may be contributing to the shift in spectral lines.

Another difficulty that has been raised by the discovery of the quasars is how to account for their immense output of energy. Measurements of their angular size by radio telescopes reveal that they are of very small dimensions compared, for example, with the size of a normal spiral galaxy. In some cases measurements indicate that the energy production is occurring in a region of space only a few light years in extent, and these conclusions from radio measurements have been supported by optical observations of the variations in light intensity. In these objects processes are occurring that produce energy equivalent to the rest mass of several million suns, but the volume of space in which this is occurring is equal to that occupied by only a few stars in a normal galaxy like the Milky Way. At present there is no known solution to this problem. Among the theoretical speculations it has been suggested that we are witnessing the annihilation of matter and antimatter. Another suggestion is that we are dealing with the collapse of a super-massive body, and that the gravitational energy released is converted into electromagnetic radiation by processes that are not entirely specified.

THE FORMATION OF GALAXIES

Although the cosmological issue remains in disarray, all plausible ideas still make the assumption that there was some primeval material from which the universe we view today has emerged. This applies even to the unchanging universe of the steady-state theory. In that case the primeval material—the hydrogen atom—is in process of continual formation. In the evolutionary group of theories, which, as already indicated, now have greater observational support, the primeval material must have been in a condition of high density 17 thousand million years ago. The assumption is made that the primeval substance must have been the simplest elementary particles as we understand them today, that is, protons and electrons. Precisely how these were arranged in the highly condensed state can only be a matter for speculation. Gamow, in developing his "hot big bang" theory, assumed that they were in the form of neutrons, and that hydrogen and other light elements were rapidly formed as the neutrons decayed into protons and electrons and thermonuclear processes began. However, it is important to emphasize that modern theory envisages the possible existence of forms of matter even more highly condensed than a condensate of neutrons.

When we considered the end of the life of a star (Chapter 3) it was seen that there is now good evidence in some cases for the collapse of the stellar material into a neutron star, where the gravitational forces override nuclear forces. If the mass of material is sufficiently great there seems no reason why further gravitational contraction should not occur, so that eventually the mass collapses inside the "event horizon" through which no signals can escape. When this consideration is applied to the singular condition of the universe at time zero in the evolutionary models of general relativity, an unaccountable condition seems to arise, in which the gravitational forces approach infinity and all matter becomes nonexistent.

Thus, at the present time the very early condition of the evolutionary universe remains obscure, and in considering the formation of galaxies and stars one can only start from the epoch when the universe consisted of a distribution of hydrogen with some helium. About 100 million years after the beginning of its expansion the mean density of this gaseous universe must have been about 10,000 times the present value, and this epoch is considered the most favorable one for the formation of galaxies from the primeval material. By processes of gravitational contraction the independent clouds of gas that eventually become the galaxies are presumed to have emerged. There is nothing in the modern theories of galaxy formation to indicate why condensations occurred in the primeval gas of the universe, or why they should have been of such a size as to form the galaxies that we find today. There is, nevertheless, one important consequence. As the galaxies condensed, their output of radiation would have heated the remaining gas, and this, coupled with the continuing expansion, would soon create conditions unfavorable for the formation of further condensations. Therefore it is believed, from these purely general considerations, that the formation of galaxies occurred in a rather short time at about 100 million years after the beginning of the expansion.

When Hubble established the extragalactic nature of the nebulae, he found that they were essentially of two types. The spherical and elliptical galaxies, exhibiting little or no structure, represented about 20 percent of the galaxies he observed. Apart from a few percent with irregular structure, the remainder were of spiral formation. He believed that there was an evolutionary sequence beginning with spherical condensations, which passed through increasing degrees of ellipticity to branch into two types of spirals, the open spirals and the barred spirals. Today we realize that this cannot have been the case—for example, the stars in the ellipticals are mostly old stars, while those in the arms of the spiral galaxies are young. Further, the discoveries of radio galaxies, quasars, and other forms reveal that the types classified by Hubble include only a partial sampling of the galaxies in the universe.

Why the universe consists of these various types of galaxy is a matter for speculation. One view is that the difference between ellipticals and spirals is essentially related to how fast the gas clouds were spinning as they condensed. Gas clouds that were spinning only slowly formed into ellipticals, while those spinning quickly spread the material out into a thin disk. Why the thin disk then took the form of a spiral galaxy, with the stars in distinct arms, then becomes a problem. A currently plausible explanation is that the gas was originally uniformly dispersed throughout the thin disk. Theoretical studies indicate that in such a disk, density waves (ripples of density) would travel out from the center in a spiral pattern, and there would be a marked tendency for stars to form in the regions of greatest gas density.

New dimensions of the problem of galaxy formation are introduced by the existence of radio galaxies, quasars, and other irregular objects whose energy output may be hundreds of times greater than that of a normal spiral galaxy. The particular problem of the quasars, where the energy source may be of small dimensions, has been discussed. As for the radio galaxies, in a large number of cases the radio contours reveal a double source of emission—the separation of the two components in the Cygnus radio galaxy is 2 minutes of arc. The existing evidence suggests that this has come about because of a violent disruptive event in the nucleus of the original galaxy. Another group of abnormal galaxies are those that resemble spiral galaxies, but with exceedingly bright nuclei (the Seyfert galaxies). In these the energy output may be 100 times greater than that of an ordinary spiral, emerging mainly from a small nucleus with a predominant output in the infrared region of the spectrum. A similar excessive energy output in the infrared is found in the nearby galaxy M82, at a distance of 13 million light years (Figure 4.12). Photographs of this object reveal filaments of hydrogen streaming outwards at a speed of 600 miles per second from an explosion estimated to have occurred 2 million years earlier in the life of the galaxy (that is, 15 million years ago in the terrestrial time frame).

As in the case of the quasars, a wide variety of suggestions have been made to explain these abnormal objects with their high energy output, often accompanied by evidence of violent events in their nuclei. Gravitational collapse, successively triggered supernova explosions, and matter-antimatter collisions are examples of the extreme hypotheses that have been invoked. No theory yet seems adequate, nor is it known whether the objects exhibiting violent events and high energy output represent any particular evolutionary phase of a galaxy. For example, one popular idea is that quasars are galaxies in early evolutionary phases, the argument being that the majority of known quasars have redshifts greater than those of most galaxies, and hence are more distant; they are therefore seen

FIGURE 4.12. *The galaxy M82, showing the results of an explosion in the nucleus. M82 is at a distance of 13 million light years, and the explosion occurred about 15 million years ago (our time frame). That is, the phogograph is of the galaxy about 2 million years after the explosive event. Motions of the streamers at a rate of 600 miles per second have been deduced. Photographed in red light by the 200-inch telescope.* (*Hale Observatories*)

at an earlier epoch of the universe. However, even if the lingering doubts about the interpretation of the quasar redshifts are dispelled, it may still be argued that the lack of normal galaxies at high redshifts is due merely to observational selection—namely, they are too faint to be seen at such great distances.

This problem is especially intractable because there seems to be no possibility of observing galaxies in process of formation—at least if the current view is correct, that their formative period occupied a limited interval in the history of the universe less than a billion years after the beginning of the expansion of the primeval material.

MAN'S RELATION TO THE UNIVERSE

At any time during the last 2,000 years individuals of several disciplines might have discoursed on man's relation to the universe. At any time during this period the claim could have been made that man was on the verge of solving two of the great intellectual problems, the origin of the universe and man's place in it. Today, the suggestion that we may be on the verge of attaining these solutions engenders a cautionary attitude. There is a feeling of humility, because of the pessimism arising from the failure of the brilliant techniques of our age, which, in every observation designed to be decisive, often seem merely to uncover ever deeper mysteries. Then there is a feeling of astonishment, because contemporary theories predict initial and final states of the universe that seem to be beyond human comprehension.

THE DEVELOPMENT OF THE CONCEPTUAL PROBLEM

Today the close involvement of these erudite studies of the universe with practical affairs has replaced the bitter intellectual and theological conflict aroused by the investigations of the universe in the 17th century. The name of Galileo is inseparably linked with that remarkable epoch, during which observational evidence undermined the belief in the perfection of the heavenly bodies and the Earth-centered, or geocentric, concept of the universe. The dispute erupted in his name because it was he who first used the newly invented telescope in a serious study of the heavens. The belief in a fixed Earth at the center of the universe could not survive his discovery of the moons of Jupiter and the phases of Venus. It was the concept of the Earth in motion, rather than the realization that heavenly bodies are not perfect, that really disturbed the contemporary theologians. Indeed, nearly 40 years before the Galilean conflict Tycho Brahe had observed the exploding star or supernova of 1572; Kepler observed another in 1604. Hence two visually brilliant examples of the imperfection of the stars in the heavens had been manifest without stimulating open conflict.

The conflict might have developed a century earlier, because it was Copernicus who really overthrew the Earth-centered static universe of Ptolemy and Aristotle. *De revolutionibus* was not printed until 1543 when Copernicus was dying. Its contents had been well known to Pope Clement VII ten years earlier; and yet it was not placed on the index of forbidden books until 1616, when Galileo had produced the observational evidence. For 73 years after the publication of Copernicus' revolutionary ideas the Church apparently did not regard them as a threat to theological doctrine.

Copernicus, Tycho Brahe, and Kepler had each presented the world with revolutionary ideas about the universe before Galileo even used his telescope. The volumes which have now been published on the trial of Galileo agree, at least, on the nature of the man as an individual of great intellectual brilliance, but arrogant and quarrelsome. It seems that the conflict was with the man who presented the new understanding of the universe. The battle was localized because those before Galileo who began to understand the nature of the universe were themselves deeply religious individuals, and for Newton, who followed him, "the true God continues to exist—always and everywhere, fashioning space and duration."

The 200 years from the beginning of the 16th to the early 18th century encompassed the genius of five men, Copernicus, Tycho Brahe, Kepler, Galileo, and Newton. In the entire history of man's understanding of the universe their individual and collective achievements remain unsurpassed. At the beginning of this epoch, the Earth was believed to be fixed at the center of the sphere of stars, and the integration of cosmological theory with one and a half thousand years of Christian dogma was complete. At its end, the Earth was known to be in motion around the Sun, and the forces that maintain it in motion were seen to apply to everything in the universe, whether a falling apple or a distant star.

It is a surprising feature of our history that after these brilliant events we did not advance our fundamental understanding of the universe for over a hundred years. Of course there were major developments in technique. Astronomers built telescopes of ever increasing size, and began to understand the physical processes occurring in the Sun and the stars. But in spite of these advances they continued to believe in a Sun-centered universe. Less than half a century ago these conceptions were only just beginning to fade in the face of the impressive observations of the American astronomers using the new 100-inch telescope on Mt. Wilson in California.

Until that time it had been firmly believed that the Sun was the center of the universe of stars that we see in the sky at night and that we call the Milky Way. Indeed there does not appear to have been much interest or thought given to the question of the extent in space of this system of stars, or what might lie beyond. The stars of the Milky Way were believed to be symmetrically arranged in space with ourselves and the Sun at the center, and they were believed to define the totality of the universe. In retrospect it is clear that the error of this belief should have been recognized a hundred years earlier. Sir William Herschel, with the amazing series of telescopes he made himself, observed over 2,000 of the objects that were called nebulae.

He suggested that these nebulae were of two distinct types: that the ones with the appearance of shining fluid were among the stars of the Milky Way, but that the nebulae apparently made up of many stars were external systems existing in space beyond the Milky Way. But no one was able to pursue Herschel's concept of external systems in a decisive manner until the decade 1920–1930.

Clarification followed the discovery of means by which stellar distances might be measured beyond the limit of parallax measurements. The observational evidence for the existence of extragalactic star systems ended the long period of uncertainty just over forty years ago. Philosophically the world had been long prepared for this observational verification. Thomas Wright, the English instrument maker, speculated on the existence of other stellar systems in his "New hypothesis of the universe," published in 1750, and five years later Immanuel Kant enunciated his theory of the island universes: "Their analogy with our own system of stars; their form, which is precisely what it should be according to our theory; the faintness of their light, which denotes an infinite distance; all are in admirable accord and lead us to consider these elliptical spots as systems of the same order as ours—in a word, to be Milky Ways similar to the one whose constitution we have explained . . . a vast field lies open to discoveries, and observation alone will give the key."

It was 174 years later that Hubble produced observational verification of Kant's philosophical conclusion. Within the space of ten years Shapley found that the Sun is an insignificant body in an undistinguished position in the Milky Way, and Hubble that the Milky Way, with its 100,000 million stars, is but an insignificant part of the universe within the observational range of the 100-inch telescope. Subsequently, the redshift measurements have been extended to ever increasing distances. Although there are many uncertainties in our current picture of the universe, it seems incontrovertible that, at least out to a distance of 4,500 million light years, galaxies have been observed that continue to follow a linear law of increasing recessional velocity with distance. At that distance the redshifts imply a velocity of recession of 46 percent of the velocity of light, that is, 85,000 miles per second. The only rational interpretation of these measurements is that the universe is expanding, and that it is doing so at an explosive rate. For example, the implication is that within the short time of 30 minutes we have separated by about 150 million miles from galaxies that can be readily photographed by modern telescopes. The fact that the concept of the expanding universe has become familiar in human language does not lessen the difficulty the human mind finds in comprehending the significance of the phenomenon.

THE IMPACT OF GENERAL RELATIVITY

In the 5th century B.C. the genius of certain Greek thinkers led them to envisage a spherical Earth. To the average citizen the idea was absurd. The Earth we, as ordinary mortals, perceive is clearly flat. Indeed, it is only in our own generation that men in high-flying aircraft and space vehicles have perceived the Earth as a spherical body. The famous paper published by Einstein in 1916 raised a similar intellectual problem in our concept of space. In the *Principia*, Newton had written of space as absolute, which "in its own nature, without regard to anything external, remains always similar and immovable." For 250 years the gravitational laws based on this postulate and on the Keplerian laws of planetary motion appeared to be verified whenever they were subject to observational test.

The considerations advanced by Einstein made it necessary to abandon the concept of space as absolute. It is, of course, possible to imagine an empty or absolute space, but this has no relevance to nature. In the universe, space has no actual significance when divorced from our measurements and perceptions of the bodies it contains. The force of gravitation in Newtonian theory is expressed simply as an empirical law, namely that the attraction between two bodies, whether they be an apple and the Earth, or the Earth and the Sun, is proportional to the product of the masses of the two bodies and to the inverse square of the distance between them. In Einstein's theory of relativity, gravitation is conceived in terms of the deformation of space near massive bodies. For Newton, space was absolute in the sense that the presence of a massive body would in no way influence the nature of the space around it. For Einstein, the properties of space are a function of the bodies contained in the universe.

Vital differences separate the concepts of Einstein and Newton. In deriving his laws of gravitation, Newton assigned to each body a gravitational mass, and it is this mass we refer to when calculating the force of attraction between two bodies. In addition, Newton's second law of motion states that the acceleration of a body—for example, an object that is being pushed along a table—depends on the force with which it is pushed and also on a property that is called its inertial mass. Newton called this the "quantity of matter" possessed by the body. Commonly it is known as the inertia of the body—its resistance to acceleration. Conceptually these two properties of a body are quite distinct. Its gravitational mass determines the force it exerts on another body, while its inertial mass is a measure of its resistance to acceleration. The functions of the two appear to be quite different, and yet Newton made experiments, by swinging pendulum bobs of different materials, that showed that the two masses were the same. Subsequently, repeated experiments with highly refined measuring systems showed that these two apparently conceptually different properties of a body were identical. In 1890, for example, Eötvös proved that the inertial and gravitational mass of a body were equal to within 1 part in 100 million. Classical physics cannot account for this identity, but it is the essence of general relativity that this identity appears as a natural consequence of the theory, because gravitational forces are manifestations of the inertia of bodies in space, which is itself modified by the presence of these bodies.

Of course, as far as normal macroscopic phenomena are concerned, the differences in the predictions of the classical Newtonian theory and general relativity are negligible, and even in some of the refined astronomical tests to which the theory of relativity has been subjected the predictions are hard to separate, so much so that there is a continuing debate about the validity and interpretation of general relativity. For example, take the case of a ray of starlight passing close to the Sun. Newton would have expected the light to continue in a straight line. But at the turn of the century, with the development of the quantum theory and the special theory of relativity, predicting the equivalence of mass and energy, a ray of light assumed the property of inertial mass. Hence, on classical Newtonian gravitational theory it may be calculated that light from a star will be deflected by nearly 1 second of arc if it grazes the Sun on its way to an observer on Earth. On Einstein's general theory of relativity the path of the ray of light from the star will curve in the solar vicinity, because of the curvature of space associated with the massive Sun. The calculated deflection in general relativity is twice that in classical theory. The effects are minute and difficult to measure, and there has been continued dispute whether the measurements made during successive total eclipses of the Sun really are in full agreement with the relativistic predictions.

General relativity raises a deep philosophical problem about the universe. The theory predicts that space-time is curved in the neighborhood of massive bodies. But does the massive body cause the curvature of space-time, or is the curvature itself responsible for the existence of matter? The theory predicts a number of small measurable local effects, such as the deflection of a ray of light passing near the Sun or the precession of the perihelion of a planet, but these can, in any case, be expressed as small corrections to Newtonian theory. The general difficulty becomes manifest when we inquire if relativity theory leads to a unique prediction about the the state of the universe.

It was natural that Einstein should seek the answer to this question during the evolution of his theory. The local properties of the universe clearly depended on the distribution of its massive bodies, as we have seen. In the latter half of the 19th century the Austrian physicist and philosopher Ernst Mach had rejected the Newtonian concept of absolute space. In his *Science of Mechanics* he argued that the behavior of any one body in the universe could only be consided in relation to all other bodies in the universe. He wrote,

"When we say that a body preserves unchanged its direction and velocity *in space*, our assertion is nothing more or less than an abbreviated reference to the *entire universe*."

In 1917 Einstein sought a generalization of this principle, that the inertial properties of matter on a small scale are determined by the behavior of matter on a cosmic scale, in terms of his theory of general relativity. He referred to the concept enunciated by Mach as *Mach's principle*, and the principle of the universal relationship of inertial masses is known today by that name. It must be pointed out, however, that the principle was firmly enunciated by Bishop Berkeley 150 years before Mach. In his *De motu*, published in 1721, Berkeley included in his general rejection of abstract ideas the Newtonian concept of absolute space and time. For him, "every place is relative, every motion relative. If all bodies are destroyed we shall be left with mere nothing, for all the attributes assigned to empty space are immediately seen to be privative or negative except its extension. But this, when space is literally empty, cannot be described or measured and so it too is effectively nothing."

It must be remembered that in 1917, the concept of the universe had none of the dynamic properties subsequently assigned to it as a result of the work of Hubble. Einstein's problem was therefore to find a unique solution of the relativity equations that would yield a static universe. He achieved success quickly. He found it necessary to introduce into the equations a new constant, the lambda term or cosmological constant, which was related in a simple way to the mean density of the universe—in fact λ was equal to the reciprocal of the square of the radius of the universe. There was also a simple relation between the total mass in the universe and the radius. Hence the universe was finite, with a radius determined by the average density assigned to the matter it contained. However, because of the curvature of space the universe was unbounded, and under certain conditions it could be circumnavigated by a ray of light.

For a short time it seemed that the general theory of relativity had yielded a unique model of the universe. It was finite yet unbounded, and it was static. There were no boundary conditions and no problem of an initial condition. The radius was simply related to the mass, and hence to the density. At that time it could perhaps have been claimed that there was no cosmological problem. It was a unique, unprecedented, but short-lived success in the story of the attempt to understand the universe.

Within a few months de Sitter showed that Einstein's solution was not unique, and using the same assumptions he constructed a different world model. But de Sitter's model had the remarkable property that it was only static if empty of matter. If a nebula were introduced into the model then it would not remain at rest, as in Einstein's solution, but would acquire an ever-increasing velocity from the observer. Five years later the Russian mathematician A. Friedmann investigated the problem without assuming that the world model was to be static. In other words he investigated the solutions of the equations in which the radius of curvature of the universe and the mean density of matter vary with time.

Friedmann's work started a line of development that at this moment seems to be without end. The observational discovery of the expansion of the universe by Hubble seven years after Friedmann's work was, naturally, hailed as a triumph for the theory. However, at this moment, over fifty years later, it is difficult to see wherein the triumph lay. Theorists of great eminence have been associated with the interpretation of the Einstein cosmology since Friedmann's work, and a multiplicity of world models have resulted. Einstein thought he had found a unique solution. The fact is that as soon as the concept of the static universe is discarded, as it must be in light of the discovery of the expansion of the universe, then an infinite set of theoretical models may be derived from general relativity. Some broad classifications of these have been discussed in Chapter 4. The simplest and most widely popularized one is that developed by G. Gamow, in which the universe originated in an explosive creation about 17,000 million years ago.

THE PROBLEM OF THE BEGINNING

The concept of a universe with a radius and a density varying with time inevitably introduces the difficulty of a superdense or singular condition at some past epoch. This difficulty is not evaded by any variations of the theory. For example, if the cosmological constant is retained and given a positive value then theoretical models of the type studied by the Abbé Lemaître are obtained. In these the initial condition of creation is pushed back 50,000 million years to a primeval atom that disintegrated; after an expansion to a size of about a million light years the primordial universe was a nearly stable configuration of hydrogen gas. Then 17,000 million years ago, condensations began to occur in this gas that eventually formed the galaxies of stars. In the Einstein equations the effect of a positive cosmological constant is that, on the large scale, a repulsive force operates in opposition to the force of Newtonian attraction, and this leads to the continual expansion of the universe. In other models, with a negative value of the constant, the universe becomes one that also exists forever, but expansion from a high-density condition alternates with contraction back to this condition, with a period that may be of the order of 100,000 million years.

Nearly fifty years ago the work of Hubble achieved a measure of agreement between observation and theory as regards the expansion of the universe. During the last few years, counts of distant radio sources and observations of microwave background radiation, described in Chapter 4, have given decisive indications of a singular high-density and high-temperature state of the universe 17 billion years ago. This presents a formidable conceptual problem. The obscurity of the issue has been mentioned. The concept of the black hole as the final state of a massive star has become familiar. With a sufficiently great mass of material the gravitational forces may override nuclear forces, so that contraction may continue until the star collapses inside the event horizon through which no signals can escape. When this consideration is applied to the singular condition of the universe in the evolutionary models of general relativity, it would seem that the gravitational forces must approach infinity, and that the entire primeval matter of the universe must have existed within a primeval black hole. The escape of any signal or event leading to the universe as it subsequently became observable is a profound problem, and inevitably raises doubts about the universal validity of physical theory as presently understood.

Indeed, it seems that modern physical theory leads to predictions concerning this problem that confound the theory itself. The prediction by general relativity of the singularity of high density at the beginning of the universe implies a near-infinite curvature of space-time. The effect of space curvature on a human body in the Sun-Earth system is negligible. However, a human being standing on the surface of a white dwarf star would feel a difference of about one-fifth his normal Earth weight between the star's gravitational attraction on his head and its attraction on his feet; at the surface of a neutron star this difference would be several million times his normal Earth weight. The physical effects as the space-time curvature approaches infinity are quite unknown. Furthermore, if the near-infinite space-time curvature of the singularity can influence the external world then a fundamental uncertainty is introduced into physical theory.

Such doubts are raised by the assumption that the singularity, or the collapse to a singular condition, must always lead to a black hole, in which all the matter lies inside the event horizon. There is no proof of this assumption. Indeed, since an extended object cannot collapse gravitationally if it is spinning rapidly, the problem concerns the possible effects of the rotation of a black hole. This is of great importance to an understanding of the earliest stage of the expansion. These conceptual problems do not arise until we inquire into the state of matter within a microsecond of the initial moment of expansion. The theories of particle physics and of quantum electrodynamics break down for space curvatures exceeding about 10^{-15} centimeters, and within the initial singularity, considered as a black hole, curvatures must be of the order 10^{-33} centimeters. The transition from unknown to known states of physics must have occurred about a microsecond after the beginning. Subsequent densities and space curvatures were not dissimilar from those accountable by nuclear physics. However, there can be no confidence in the applicability at earlier times of the laws of physics as they are at present known to us on Earth.

There is one observed feature of the universe when viewed on the large scale—the high degree of uniformity or isotropy—that may have been conditioned by the extreme physical conditions, especially of the interaction of matter and radiation, within the first microseconds of the beginning. These epochs of space-time are forever beyond the powers of observation, and it is not unreasonable to feel that they may also lie beyond the scope of physical theory and knowledge. Indeed, on the most pessimistic view it could be maintained that we cannot investigate observationally, and possibly not theoretically, the crucial epoch when the most striking feature of our universe was determined—its large-scale uniformity.

THE FUTURE HISTORY OF THE UNIVERSE

Whether or not the initial state is subject to physical inquiry, the available observational evidence leads to the concept that the universe is expanding from a singular condition of high density. What will ultimately happen to the universe? In the present state of our knowledge there would appear to be two main possibilities, which it would be possible to assess more realistically if the deceleration parameter q, discussed in Chapter 4, could be determined. If q is equal to or less than $\frac{1}{2}$ the universe will expand indefinitely. More and more of the galaxies will move over the observational horizon, the energy of the universe will become less and less available, and eventually the universe will move towards what is conventionally called a heat death. But if q is greater than $\frac{1}{2}$ then the expansion will at some future time reach a maximum, and the universe will contract to another singular condition of high density. In its simplest form the problem is whether the density of the matter in the universe is sufficient to overcome, by self-gravitation, the forces that are causing the expansion. The critical density is estimated to be 2×10^{-29} grams per cubic centimeter. The matter in galaxies is believed to supply only a few percent of this critical value. If the universe is to contract again, there must be large amounts of matter in intergalactic space, or other objects that have yet to be discovered. The exact definition of the Hubble line at large redshifts could settle the problem unambiguously. Indeed there are some cosmologists who already believe that the available evidence supports a deceleration parameter greater than $\frac{1}{2}$, and hence a future contracting phase.

The concept of a universe behaving in a way indicated by Figure 4.6 leads to interesting speculations about the future. We exist in an expanding universe, which is to say the scale factor R is increasing. If the deceleration parameter q is greater than $\frac{1}{2}$, this expansion will eventually cease and the universe will begin to collapse. The time to the change from expansion to contraction depends on the value of q. If, for example, $q = 1$ then we exist in an epoch at which about one-tenth of the cycle has elapsed, and the turning point will be reached in another 10^{10} years. If q is greater than $\frac{1}{2}$ but smaller than unity then this time will be increased. As the universe contracts the redshift of the nearby galaxies will change to blueshifts and eventually distant objects will be observed as blueshifted. The time scale of contraction cannot be predicted—it is not clear whether the expanding and contracting phases must be symmetrical. Eventually the galaxies will merge, but the stars are unlikely to meet their end by collision. Before the density increases to the extent that collisions are frequent it seems likely that the stars will be destroyed under the influence of the external radiation, which will become hotter than the stellar interiors. Since the state of matter is unknown in the singular condition, it cannot be predicted whether the loop will be closed, or whether a "bounce" will occur, and the cycle of expansion and contraction repeat indefinitely.

THE PROBLEM OF THE UNIQUENESS OF THE OBSERVABLE UNIVERSE

If the universe is cyclical, with an infinite past set of singularities and an infinite future set of loops of expansion and contraction, then the problem of a beginning and an end is evaded. However, there is no reason to believe that the physical laws and fundamental constants we find in our present cycle will have occurred previously or will be repeated in a future cycle. At least some cosmologists take the view that we exist in this particular cycle because the laws of physics and the fundamental constants are consistent with the emergence of biological matter.

A clear example of the delicacy of the balance in the early history of the universe was pointed out a few years ago by Dyson.* Why does the Sun burn its hydrogen in the thermonuclear transformation to helium instead of blowing up explosively like the hydrogen bomb? The answer lies in the nature of the forces between particles when they approach to within about 10^{-13} centimeters. These are the extremely strong forces that bind protons and neutrons in the atomic nucleus. However, not all nuclear reactions exhibit this powerful coupling. For example, in the Sun the basic transformation process involves ordinary hydrogen and the fusion of protons

$$\text{proton} + \text{proton} \rightarrow \text{deuteron} + \text{positron} + \text{neutrino}$$

whereas in the bomb the transformation process involves the heavy hydrogen isotopes, deuterium and tritium, and the fusion of deuterons

$$\text{deuteron} + \text{deuteron} \rightarrow \text{helium-3} + \text{neutron}$$

Now this second process ($d + d$) is a strong nuclear interaction, whereas the solar reaction ($p + p$) is a weak nuclear interaction. The consequent difference in the speed of the reactions is immense—a factor of about 10^{18}. From a purely terrestrial point of view the weakness of the $p + p$ reaction is fortunate, because it means that the Sun produces its energy over a period of billions of years, rather than explosively. Incidentally, the weakness of the reaction is one of the fundamental reasons why the oceans cannot be turned into gigantic nuclear bombs.

The existence of the universe as we know it depends on a far more extraordinary circumstance. In principle the interaction of two protons can yield the helium-2 isotope as follows:

$$\text{proton} + \text{proton} \rightarrow \text{helium-2} + \text{photon}$$

---

*F. J. Dyson, *Scientific American 225*, 51, 1971.

The helium-2 nucleus consists of two protons and no neutrons, and decays spontaneously into a deuteron thus:

$$\text{helium-2} \rightarrow \text{deuteron} + \text{positron} + \text{neutrino}$$

This proton fusion could proceed as a strong interaction but the coupling is a little too weak to produce a bound state of helium-2. Although the helium-2 state is well observed, it is unstable by only half a million volts. Since the $(p + p)$ attractive force is about 20 million volts, the difference is only about $2\frac{1}{2}$ percent. In other words, if the proton-proton interaction had been only a few percent stronger, all the primeval hydrogen of the universe would have been synthesized into helium in the early stages of the expansion. This would have happened before the galaxies formed. No long-lived stars like the Sun would have been born, and life as we know it would never have emerged.

The possibility that the universe is cyclic also raises puzzling questions about the direction, or arrow, of time. We have clear concepts in our terrestrial life of the direction of this arrow and the meaning of "before" and "after." There is continuing debate about the arrow of time as far as the universe is concerned. If the arrow is in the sense of increasing time in this expanding phase, what happens at the turning point of the cyclic universe? Does time's arrow reverse as the contraction begins? Our whole concept of time and of before and after may be too naive, and inapplicable to the extreme conditions of the singularities. Indeed, at least one eminent cosmologist, J. A. Wheeler of Princeton, has suggested that in successive cycles of the universe the emerging physical laws are governed by some process of natural selection. In some cycles of the universe, like our present one, the fundamental constants may be such as to facilitate the condensation of stars with long-term stable conditions, and hence the emergence of biological material. In other cycles the conditions may be inimical to life and remain forever unknowable. Man's relation to the universe, and indeed the possibility of his emergence, seems to have been determined by events in the first microsecond of time, seventeen billion years ago. Whether the isotropy, which seems to be a prime requirement for our existence, was itself determined in the initial state, or whether it is an inevitable consequence of a wide range of initial states, remains uncertain. The observational and theoretical studies in cosmology of recent years indicate a remarkable and essential relationship between our existence, the fundamental constants of nature, and the initial moments of space and time.

INDEX